100
FBBA
AF678025

Sport in the Global Society

General Editors: J.A. Mangan and Boria Majumdar

LEISURE AND RECREATION IN A VICTORIAN MINING COMMUNITY

Sport in the Global Society

General Editors: J.A. Mangan and Boria Majumdar

The interest in sports studies around the world is growing and will continue to do so. This unique series combines aspects of the expanding study of *sport in the global society*, providing comprehensiveness and comparison under one editorial umbrella. It is particularly timely, with studies in the aesthetic elements of sport proliferating in institutions of higher education.

Eric Hobsbawm once called sport one of the most significant practices of the late nineteenth century. Its significance was even more marked in the late twentieth century and will continue to grow in importance into the new millennium as the world develops into a 'global village' sharing the English language, technology and sport.

Other Titles in the Series

Disreputable Pleasures
Less Virtuous Victorians at Play
Edited by Mike Huggins and J.A. Mangan

Italian Fascism and the Female Body
Sport, Submissive Women and Strong Mothers
Gigliola Gori

Rugby's Great Split
Class, Culture and the Origins of Rugby League Football
Tony Collins

Terrace Heroes
The Life and Times of the 1930s Professional Footballer
Graham Kelly

Barbarians, Gentlemen and Players
A Sociological Study of the Development of Rugby Football
Second Edition
Eric Dunning and Kenneth Sheard

Capoeira
The History of an Afro-Brazilian Martial Art
Matthias Röhrig Assunção

British Football and Social Exclusion
Edited by Stephen Wagg

The First Black Footballer
Arthur Wharton 1865–1930: An Absence of Memory
Phil Vasili

Leisure and Recreation in a Victorian Mining Community

Leisure and Recreation in a Victorian Mining Community describes a working-class mining community in the industrial era, at play. This detailed historical study provides a fascinating perspective on a way of life more usually characterized by its economic hardship and social rigidity than by its pastimes and pleasures.

The text addresses working-class sport and leisure in its totality, using a range of approaches to explore the life and history of this clearly defined period and geographical location. With a full introduction describing the social context of these communities in East Northumberland, the text goes on to analyse:

- The transformation of traditional sports
- The impact of "new sports"
- Non-sporting leisure
- Sport and leisure, social change and social identity

All are placed within the context of other working-class communities in Britain and a case study of football draws together the key themes of social change, tradition, identity and social meaning that run throughout the book.

The product of many years' research, this unique study reveals a complex social reality, and represents an important resource for all historians of society, sport and leisure.

Alan Metcalfe was born in Jarrow, England. He is Emeritus Professor of Kinesiology at the University of Windsor and has published widely on Canadian and British sports history.

Leisure and Recreation in a Victorian Mining Community

The Social Economy of Leisure in North-East England, 1820–1914

Alan Metcalfe

Routledge
Taylor & Francis Group
LONDON AND NEW YORK

First published 2006
by Routledge
2 Park Square, Milton Park, Abingdon, Oxon OX14 4RN

Simultaneously published in the USA and Canada
by Routledge
270 Madison Ave, New York, NY 10016

Routledge is an imprint of the Taylor & Francis Group

© 2006 Alan Metcalfe

Typeset in Baskerville by
Keystroke, Jacaranda Lodge, Wolverhampton
Printed and bound in Great Britain by
Antony Rowe Ltd, Chippenham, Wiltshire

All rights reserved. No part of this book may be reprinted or reproduced or utilised in any form or by any electronic, mechanical, or other means, now known or hereafter invented, including photocopying and recording, or in any information storage or retrieval system, without permission in writing from the publishers.

British Library Cataloguing in Publication Data
A catalogue record for this book is available from the British Library

Library of Congress Cataloging in Publication Data
Metcalfe, Alan.
Leisure and recreation in a Victorian mining community : the social economy of leisure in rural north-east England, 1820–1914 / Alan Metcalfe.
p. cm. – (Sport in the global society)
Includes bibliographical references and index.
ISBN 0–415–35697–0 (hardback) – ISBN 0–203–00296–2 (ebook)
1. Working class–Recreation–England–Northumberland–History–19th century. 2. Miners–Recreation–England–Northumberland–History–19th century. 3. Leisure–England–Northumberland–History–19th century. 4. Recreation–England–Northumberland–History–19th century. I. Title. II. Series.
GV76 .E6N67 2005
790 .1′086′23094288–dc22 2005005969

ISBN10 0–415–35697–0 ISBN13 9–78–0–415–35697–8

For David Starbuck and my two granddaughters Kaitlin and Mckenna

Contents

List of plates

Series editors' foreword

This is a significant book about working people 'significantly different from other working people' – the coal-miners.[1]

William J. Bouwsma declared in the 1980s that history is parasitical. As it changes so does the host on which it feeds; intellectual history, once nourished by philosophy, thus now feeds off the arts. He expected this appetite to grow stronger and to consume popular, as well as elite, artistic expression including music and dance.[2] He omitted to mention sport, a singular omission. Alan Metcalfe makes no such mistake. He has written a lasting contribution to the local history of working-class sport. But this is not all; he has brought the leisure and recreational pastimes of the East Northumberland miner out of the "underground" shadows and into bright sunlight.

He demonstrates the truth of the often repeated maxim: men make their own history but not in conditions of their own making. For the nineteenth-century miner those conditions were horrific, "the long losing battle with poverty, under-nourishment and insecurity, . . . the risk of accident and maiming, . . . the rot of body and soul and the dread of the workhouse at the end of that bitter road."[3]

Despite such conditions, by the 1860s what was true of the "first industrial society" of Lancashire – working men forming clubs of their own, often in the face of considerable difficulties with the dyers, finishers and bleachers of Middleton, for example, playing cricket on a "ground" twenty-six yards by twelve with a brook on one side[4] – was equally true of the coal-mining community of East Northumberland.

Metcalfe, however, has dug deep and exposed a rich seam of activities, "that covered the whole (recreational) spectrum from Shaw to Ibsen to melodramas, symphonic music to brass bands, lawn tennis to potshare bowling."[5] He goes one better than Bernice Martin, who remarked of her consideration of cultural change (in the post-Second World War decades) that, while she based her case on England, in her opinion, it had validity for the modern Western world, but nevertheless she had too much respect for historical and geographical specificities to assume that the processes she delineated for England were exactly paralleled, even in such broadly comparable societies. She concluded that "each case needs to be separately established and documented before reliable comparisons can be made.[6] Metcalfe rightly beats his metaphorical drum even more vigorously. In his measured view, there was no sense of an overriding [recreational] working class culture among the miners. Differences were repeatedly greater than similarities. Indeed, "differences were so great that villages only a mile away from each other had significantly different sporting histories – some had none."[7] He is also insistent on the co-existence of continuity and discontinuity, echoing Martin Weiner's comment: "Victorian achievements were Janus-faced. If society was transformed . . . the extent of the transformation was more united than it first appeared to be. . . . New forces did not tear the social fabric. Old values lived on with the new . . . the end result was the nineteenth-century transformation that was a peaceful accommodation."[8] The recreation of the East Northumberland miner saw change: from innkeeper sponsor, from public land to enclosed land competitions, from amateur to "professional" organizer and from local to regional and national jurisdiction, while at the same time determinedly independent social clubs maintained interest in traditional activities.

This grassroots story has an impressive subtlety which takes careful account of attempts at cultural stability in a time of rapid change and local variation within a "homogeneous" community.

J.A. Mangan
Boria Majumdar
Series Editors
Sport in the Global Society

Acknowledgements

It is extremely difficult to identify who has influenced the writing of a book that had its academic origins nearly 40 years ago. Nearly by definition the acknowledgements will be incomplete mainly because I have difficulty recognizing the ways in which my thinking changed over time. However, the book is deeply rooted in my past, in particular, my childhood years on Tyneside. I have remained a Geordie throughout, despite living in Canada for 45 years. As an academic study it is rooted in my graduate work at the University of Wisconsin (1965–68). It is very much influenced by two professors: J.F.C. Harrison, the eminent English historian; and G. Kenyon, my adviser in the Faculty of Physical Education. Without them this book would never have been written. However, it was from within the emerging field of sport and physical education history that the major influences emerged. Four early leaders placed an imprint on sport history and me: the late Bruce Bennett, Marvin Eyler, Max Howell and Earle Zeigler. Many of the ideas underlying this book were honed in discussions at the annual meetings of the North American Society for Sport History. It was at NASSH that I presented my ideas and received feedback. Among all the people with whom I interacted there is a group that I now recognize as contributing to the development of my ideas. They are, in alphabetical order, Mel Adelman, Douglas Booth, Nancy Bouchier, Dick Crepeau, Larry Fielding, Rick Gruneau, Steve Hardy, Bruce Kidd, "Chuck" Korr, Don Morrow, Vicki Paraschak, Tina Parratt, Mike Salter, Nancy Struna, and Pat Vertinsky. The works of these colleagues, with another group of historians, have inspired me. The others include Larry Gerlach, Allen Guttmann, Colin Howell, Arnd Krueger, Roberta Park, Steven Riess, Ben Rader, Roland Renson, Ron Smith, Horst Ueberhorst, the late A.J. "Sandy" Young and David Zang.

Of course, this is very much located in England and, in particular the North East of England. For primary source material I am deeply indebted to the professional staff of the Local Studies Department of Newcastle Public Library, especially in the late 1980s and early 1990s. Similarly the staff at the Northumberland County Record Offices at Melton Park and Morpeth were always courteous and helpful. I owe an immense debt of gratitude to my life-long friend David Starbuck who acted as my Newcastle based "gopher". He retrieved large amounts of material and chapter 2 is based entirely on his research activities. Some work produced by a small group of local historians has proved invaluable. Pre-eminent amongst these is the late Stephen Martin, whose booklets on the mining villages of Bedlingtonshire provide a treasure trove of information and insights into mining life. His work has been extended by his son Evan Martin. Similar work has been done by Bob Balmer and the Blyth Local Studies Committee on Blyth, Mike Kirkup on Ashington, and Thomas Allan on Seaton Delaval.

This primary data was placed within an academic context provided by some North East historians and some sport historians. In terms of the North East four names loom large in terms of this work: Norman McCord, Robert Colls, Bill Lancaster and Bill Williamson. With reference to British sport history, in particular working-class sport history, the works of Jeff Hill, Richard Holt, Tony Mason, Neil Tranter and Jack Williams have proved to be invaluable.

Of course, as any author recognizes, getting the manuscript to a publisher is only the first step. My very rough manuscript was first read by Richard Holt and Jeff Hill. Their incisive comments proved to be a great help. However, any merit the book has is due to the outstanding editorial work of Tony Mangan and Jonathan Manley. In fact, the review process put in place by Tony proved to be one of the most worthwhile academic experiences of my life. I owe an immense debt of gratitude to Tony Mangan and the two reviewers – thank you whoever you are. Finally, I am indebted to Kate Manson and Liza Mackenzie who entered the game at a late stage but had a major impact upon the process.

I would be remiss if I did not express my gratitude to the University of Windsor and, in particular, the Department of Kinesiology for their support throughout the past 36 years. They have allowed me to pursue this somewhat esoteric topic.

Finally my family has been involved in the whole process and has shown remarkable restraint as I pontificated about the miners of East Northumberland. In fact, after many visits to Newcastle my two daughters Alisa and Karen have become honorary Geordies and retain their love of the North East. As for my wife, Heather, I really don't know how she put up with my fixation on the mining villages of East Northumberland. She is, quite simply, the driving force behind everything I do. There are no words that can adequately express my debt to her.

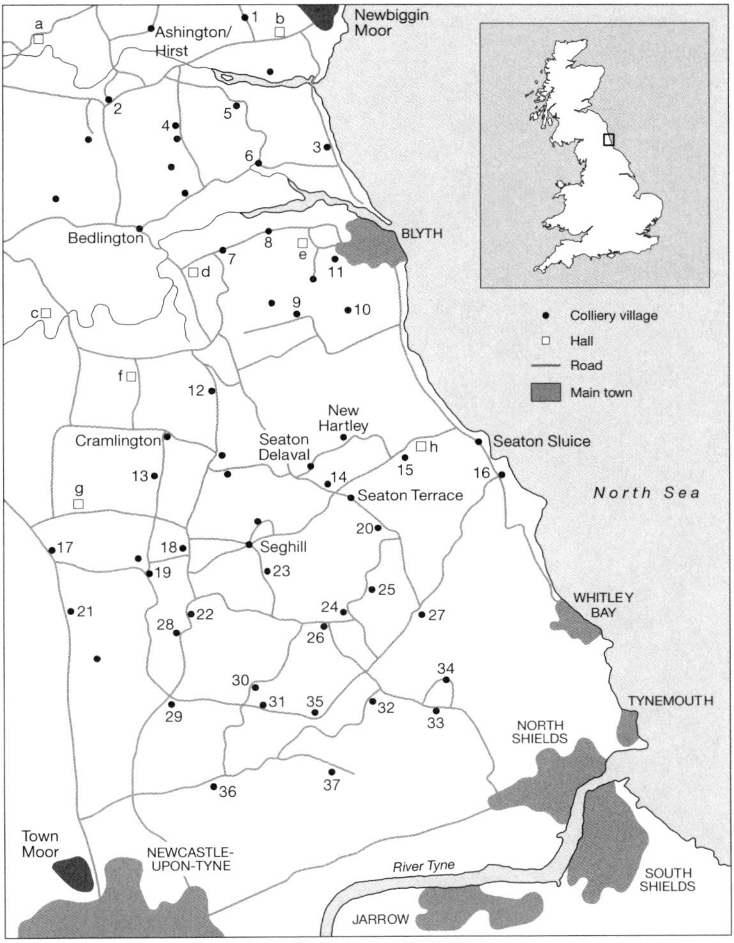

Adapted from Craster, H.H.E (1909) *A History of Northumberland*, Vol. IX

The colliery villages of Victorian East Northumberland.

Key to Map

Mining villages

1. Seaton Hirst
2. Guide Post
3. Cambois
4. Bomarsund
5. West Sleekburn
6. East Sleekburn
7. Bebside
8. Cowpen
9. New Delaval
10. Newsham
11. Waterloo
12. Shankhouse
13. West Cramlington
14. Avenue Head
15. Seaton
16. Hartley
17. Seaton Burn
18. Annitsford
19. Dudley
20. Holywell
21. Wideopen
22. Burradon
23. Backworth Colliery
24. West Holywell
25. East Holywell
26. Backworth
27. Earsdon
28. Camperdown
29. West Moor
30. Killingworth
31. Palmersville
32. The Allotment
33. New York
34. Murton
35. Benton Square
36. Longbenton
37. Rising Sun

Halls

a. Bothal Castle
b. North Seaton Hall
c. Hartford Hall
d. Bebside Hall
e. Cowpen Hall
f. Crow Hall
g. Arcot Hall
h. Delaval Hall

Introduction

No book exists in a vacuum. It is written within the context of the social, intellectual and cultural background of the writer. In fact, the roots of this monograph lie deep in my past. My place of birth had a major impact upon my academic interests. I was born in Jarrow-upon-Tyne in 1933 in the midst of the Depression. Throughout my 17 years in Jarrow the town never lost an air of depression. Men hunkering down on the street corners, a sense of alienation and depression, a general feeling of inferiority and a slowly decaying town was the environment within which I spent my formative years. It was above anything else a working-class town. In the Lower Sixth at Jarrow Grammar School I met Bob Ranson. His father was a coal miner. Visiting the Ranson house I moved into a totally new world, a world alien to anything I had ever experienced, the world of a coal miner's family. Even then I recognized that miners were different: Mr Ranson exuded an enthusiasm, a sense of self, and an independence that I had never experienced and it was a sense of pride and self-confidence that made Mr Ranson so different from other working-class men that I had met. This sense of difference was reinforced when playing for the Marine School 4th XV rugby team against Seghill, a mining village in East Northumberland. I had never experienced such ferocious physicality on the playing field. In somewhat of a paradox the social interaction after the match in the Seghill clubhouse was amongst the most pleasant and memorable that I ever experienced. Thus, even at this early stage in my life, my interest was aroused about this group of working people who were significantly different from other working people.

This perception of difference is important. There is little doubt that miners were perceived, by outsiders, to be different. The majority of their contemporaries would have agreed with the statement that, "To many Victorian observers coal-miners represented a strange, almost sub-human, stratum of the working class, pugnacious, brutalized by their grimy toil, inhabiting isolated communities which, in leisure time, became dens of drunkenness and savage sports."[1] However, to other observers the differences lay in the ways they dealt with the harshness of their living conditions. As Elfin stated in his weekly column in 1909,

> At a miners' meeting in a Northumbrian pit village "You get, in such a meeting, a fund of character, half serious, half humorous, which is to be found

> nowhere else. There is a sort of rough, unpolished intelligence in a gathering of typical Northumberland miners that speaks eloquently of persistent, praiseworthy effort at self-education, and betrays a knowledge far beyond that of the average artisan in the city. At the meeting to which I am referring, the chairman, a miner, quoted Shakespeare as a familiar friend, with whose works he was well acquainted."[2]

It is the exploration of the ways in which miners and their families strove to create meaningful lives in what can only be described as horrendous working and living conditions that lies at the heart of this monograph.

While personal interest is an important and necessary ingredient in the development of a long-term project, its importance and significance must lie on a more solid academic foundation. In order to achieve this two questions have to be addressed. First, what is the relevance of a book on working-class leisure? Secondly, what is the significance of a local history?

One element of working-class life that remains in the shadows is leisure. For the most part historians have ignored leisure and thus provided a biased and incomplete view of working-class life and the ways in which working people struggled to bring some light and meaning to life. When some social historians eventually did focus on leisure it was on particular forms of leisure, those activities falling within the boundaries of "Rational Recreation." In fact, the majority of social historians, when faced with sport and recreation, would feel as Norman McCord, the pre-eminent historian of the North East, did when he stated:

> It may seem frivolous in a book of this kind to devote so much space (5 pages) to a consideration of sport and recreation, and it would be so if for many thousands who engaged in such activities they represented merely a trivial accompaniment to a life in which the highest priorities were directed to the problems of society as a whole. It was not so, however, for it is plain that sport and recreation occupied a central place in the preoccupations of very many people within the region.[3]

McCord's statement reflects both the reluctance and the discomfort academics feel when dealing with such, to the academic mind, trivial concerns and the reality, on the other hand, of the importance of these "trivial" things to the people whose lives the activities were a part of. Historians have focused on things that they are interested in: education, religion, temperance, unions, etc. – all, as we shall see, the preserve of relatively small numbers of people. Were these the only avenues open to the uneducated majority in their pursuit of some meaning in life? What did people living in small overcrowded houses with limited facilities do in order to provide some meaning in life? They found release in a large number of activities participated in during those times away from the mine – leisure activities and sport were not peripheral to life but central. In fact, the importance of leisure was recognized, at the time, by both management and miners. Speaking of the 1850s, G. Halliwell, a miner, stated, "Amusements they must have, or life would hardly be worth living."[4] In August 1907, Professor J.H. Merivale, speaking to the Annual Meeting of the North of England Institute of Mining and Mechanical Engineers stated: "Recreation is a matter of some importance and it appeared to him that the question of how the miner would use his spare time was of very practical interest to the coal trade."[5] Sport was not just a diversion from the more important things in life but to many it was *the* most important thing in life. I suggest that for generations of miners and their families, living in conditions not conducive to the Victorian ideal of family life, it was the rich and complex set of leisure and sporting activities that provided a central focus to the lives of the men because, without any doubt, leisure and sport were male domains. This is an important caveat. Mining society was clearly divided along gender lines. Whether women had any leisure is, indeed, a moot point although fragmentary evidence suggests moments taken throughout the day.[6] However, it is even more difficult to construct a picture of women's lives than one of men's and thus, for the most part, women will be peripheral to this account.

When I embarked on this project in 1966, virtually nothing had been written on sport and leisure. For the most part, the history of leisure and the history of sport followed different paths. The foundations of leisure research were laid in the late 1970s – Meller (1976), Lowerson and Myerscough (1977), the influential book by Bailey (1978), Walvin (1978) and Cunningham (1980, 1985, 1990).[7] With respect to sport history, it developed at roughly the same time although it covered particular topics in particular ways. Mangan's *Athleticism in the Victorian and Edwardian Public School* (1981) articulated the characteristics of the dominant hegemony.[8] The emphasis upon the power of the upper classes was articulated even more clearly in Birley's work.[9] The spread of the Public School ethos, and the ideology of amateurism more deeply into the urban middle classes, was the subject of Lowerson's book in 1993.[10] These works served to articulate clearly, and in some depth, the nature and extent of the ideology of amateurism and, in so doing, provided a particular view of British sport history. Another focus has been on the histories of individual sports placed within the context of the society; Mason on football, Vamplew and Huggins on horse racing, Halladay and Wigglesworth on rowing, and Birley on cricket, to name but a few.[11] The other major focus flowed from three books which attempted to bring some focus to British sport history: Mason's *Sport in Britain* (1988); Holt's *Sport and the British* (1989); and Wigglesworth's *The Evolution of British Sport* (1996).[12] One of the major consequences of Holt's book was his recognition of "areas of patchy coverage and unexplored topics" which provided an agenda for the future.[13] One of the major areas was working-class sport history. This resulted in his edited volume of essays *Sport and the Working Class in Modern Britain* (1990).[14] Much of the subsequent work on the working class can be directly related to Holt's edited volume and it was questions raised by Holt that led to this work and provided the thematic focus to the book. One of his objectives was to uncover the "submerged tradition of popular recreation." Thus, one of the central organizing themes relates to traditional sports and the impact of the massive changes of the second half of the century. Additionally, he attacked a simplistic class analysis. Thus, a second theme of the book relates to the complexity of the relationships between various social groups. In a different vein another offshoot of Holt's book was Hill and Williams' edited volume *Sport and Identity in the North of England.* (1996).[15] These essays addressed the central question of the meaning of sport. And thus, a third thread relates to the meaning of sport to the miners of East Northumberland. However, none of these books addresses the complexity of working-class leisure and sport in its totality. And it is at this juncture that this book is different. This book attempts to integrate these different approaches to working-class sport and leisure. While most of the topics have been addressed by other historians none has attempted to examine the totality. It takes several approaches that are new. It focuses on 22 sports in a clearly defined geographical location. Only Tranter's work on Central Scotland has tried to do this.[16] Second, it addresses all social groups and the relationships between them. Finally, it looks at non-sporting leisure. This is important because the two views, sport and non-sporting leisure, provide totally different views of the community and the social relationships within it. It is these three differences that make this work significant and important for social historians, leisure historians and sport historians.

Why is this local history an important topic? It is significant to several different groups. First, for historians of the North East it fills gaps in our knowledge about this very important group of workers – a group that was one of the foundation stones of the economic success of the North East in the nineteenth century. Additionally, it is important to rescue the miners from oblivion and the "condescension of posterity." This is even more urgent since they are, as a distinct group, fast disappearing. Only one of the 66 collieries that dominated the landscape of East Northumberland in the nineteenth century remains today. However, it is also relevant to social historians, sport historians and leisure historians. Each chapter places leisure activities of the miners of East Northumberland in the wider context. First by comparing their activities with those of other miners throughout Britain and second, by looking at the similarities and differences between the miners and other working-class enclaves throughout Britain. Thus, this case study provides a view of working-class leisure and sport throughout Britain from 1820 to 1914.

Miners created their own lives, not under conditions of their own choosing but under structural constraints, historically created and transmitted from the past.[17] Therefore, chapter 1 will examine the relationship between leisure activities and the physical, social, intellectual and economic structures that influenced the lives of the inhabitants of the villages. At the heart of these lie the complex social relationships that under-girded village life. More particular to daily life were the relationships between three social groups: the colliery managers and officials; the small group of professionals – teachers, ministers and doctors; and the miners. At the same time, it is important to emphasize that the miners were not a single homogeneous group. Chapter 2 examines the non-sporting leisure activities through a case study of one village, Seaton Delaval, from 1838 down to the First World War. This is important because it provides a benchmark against which to judge the sporting activities. The examination of the sporting activities is given some focus by highlighting the changes to sport that occurred during the century – in particular, the emergence of modern, organized sport.[18] Chapter 3 focuses on some traditional sports that remained relatively unchanged throughout the course of the century. Chapter 4 examines a group of traditional sports that were changed in fundamental ways. This is related to the degree to which they adopted the characteristics of modern sport. Chapter 5 deals with sports that were truly new to the mining villages. Chapter 6, a case study of football, brings all the different elements together. Finally, a brief conclusion looks, first, at some insights into fundamental aspects of what happened during the nineteenth century and then considers the similarities and differences between East Northumberland and other mining villages and working-class enclaves.

This monograph is, therefore, a history of the leisure and sporting activities in the mining villages and towns of East Northumberland placed within the context of certain structural variables. It attempts to examine the degree to which the inhabitants of the villages created their own lives and the degree to which they were subject to the arbitrary actions of the colliery owners and officials. The view that emerges is not a simple one. What it reveals is the immense complexity of these, to some observers, apparently homogeneous communities. Leisure played an important part in providing some meaning to generations of miners and their families in conditions that, to outsiders at the time, as well as ourselves today, were horrendous.

1 The social context

East Northumberland, 1820–1914

Leisure practices do not exist in isolation, but are related to certain structural factors. Leisure activities are, and always were, constrained by the availability of time, space and money. In East Northumberland in the nineteenth and early twentieth centuries, these three elements were related in turn to decision-making power, that is the power to determine the conditions within which the miners and their families lived, worked and played. This chapter will examine these power relationships and how they influenced daily life in the villages. This will be done, first, by a consideration of decision making over land usage; second, by an examination of the mining villages, their inhabitants, their places of origin, the quality of housing, sanitation, and the necessities of life available to them; and third, by a discussion of the degree to which the aforementioned necessary conditions of leisure, time, space and money were present at different times during the period. A final section will place the mining villages of East Northumberland within the context of other working-class enclaves in Britain.

Decision making over land usage

At the beginning of the nineteenth century East Northumberland, a remote, isolated farming area with a population of 6,270, was dominated by the estates of Sir Matthew White Ridley, the Delavals, the Duke of Portland, the Duke of Northumberland, Viscount Barrington, the Earl of Carlisle, Sir Francis Blake and the Ecclesiastical Commissioners. These landowners played pivotal roles in the construction of the physical, economic and social environment of the mining villages created after the 1820s. The history of the Ridleys of Blagdon Hall illustrates the ways in which landowners influenced life within the district throughout the nineteenth and early twentieth centuries.

The Ridleys had a long history in Northumberland, but it was not until 1735 that they moved from Heaton to their new estate at Blagdon.[1] During the nineteenth century nothing could be built on their land without their permission. The construction of mining villages such as Cambois, Newsham and Cowpen was done under the control of the Ridleys. Their absolute power was eroded on March 16, 1863 when South Blyth and Cowpen adopted the Local Government Act. Power was now shared with a Local Board consisting of professionals and businessmen. This division of power soon led to disagreements between the two parties. In most instances the disputes were over control of land. Between 1875 and 1879 there was an ongoing acrimonious dispute over access to Blyth Links, a traditional location of popular recreation. Similarly between 1897 and 1904, they engaged in prolonged discussions over the donation, by Viscount Ridley, of seven acres for a public park. In each case agreement was reached only after the Council agreed to the terms laid down by the Ridleys.

The Ridleys were also involved in the economic development of the area. As early as 1788, Sir Matthew White Ridley constructed a lighthouse at the entrance to the River Blyth. They were also involved in the development of the new deep mines on the banks of the River Blyth. Later, during the 1850s, they built gas-works and waterworks to serve South Blyth and Cowpen. Perhaps their most important contribution to the economic development of the area lay in the key role of Sir Matthew White Ridley in the creation of Blyth Harbour and Dock Company in 1853.[2] This was the beginning of the development of Blyth as a major trans-shipment point for coal and the development of a small shipbuilding and repair industry. Thus, throughout the century, the Ridleys were important figures in the economic development of the area.

With respect to daily life, they donated facilities, dispensed justice and were involved in local government. From 1883, sites were donated for cricket clubs, churches, hospitals, miners' halls and Mechanics' Institutes. Successive Sir Matthew White Ridleys served at Bedlington Petty Sessions, on the bench at the Northumberland Assizes in Newcastle, and at the annual Brewster Sessions at Bedlington and Blyth. After the creation of Northumberland County Council in 1888, they served on the Council, often in the role of chairman. Simply through involvement in providing sites for building, investment in business, and involvement in justice and local government the Ridleys' influence reached into all aspects of life in East Northumberland.

In one way the Ridleys were atypical; they spent a significant amount of time at Blagdon Hall. The other major landowners, in particular, Lord Hastings at Seaton Delaval Hall and the Duke of Portland at Bothal Castle, maintained control through resident agents such as the Samples of Bothal Castle. The Samples, who came to Bothal Castle in the 1820s, had a more direct impact upon the inhabitants of the Duke's villages than the Duke himself.[3] They became part of a new social elite that emerged during the first half of the nineteenth century and moved into the newly built halls and houses. The first of these, Hartford House, was built by the Burdons in 1807. In 1824, the Lawsons built Longhirst Hall and, in the same decade, the Shum-Storeys constructed Arcot Hall near Cramlington. By the late 1850s these halls were occupied by a new elite comprising the new wealth coal owners.[4] In 1858, Edward Potter Esq., owner of Cramlington Collieries, lived in Cramlington House. Hugh Taylor Esq. J.P., one of the most successful of the new breed, lived in Backworth House. George Baker Forster and Charles Carr occupied East Cramlington House and Seghill House. Like their predecessors they, too, did not stay long in East Northumberland. As they acquired the hallmarks of respectability, education at one of the prestigious public schools, attendance at

Plate 1.1 Hartford Hall, home of the Burdons. Built in 1807.
Reprinted with permission from Martin, E. (1997) *Images of England: Bedlingtonshire*, Stroud: Tempus Publishing

Oxford or Cambridge, and membership in the appropriate London clubs, they aspired to greater social heights, heights not to be found in East Northumberland. By the turn of the century the Milburns of Ashington Coal Company had moved to Shabden Heath, Surrey; the Laycocks of Seghill Coal Company to Wiseden, Nottingham; the Taylors of Backworth Coal Company to Chipchase Castle on the North Tyne, and the Shum-Storeys of Arcot Hall to Southampton.[5] They were replaced by another "new" social elite – the agents, viewers and managers of the collieries and the land agents of the landed aristocracy. By 1902 the sole surviving member of the old landed aristocracy was Viscount Ridley at Blagdon Hall. There were two other old Northumbrian families, the Burdons at Hartford House and the Sidneys at Cowpen Hall. They were joined by T.E. Jobling, colliery agent and viewer, in Bebside Hall; Robert Embleton Ornsby, agent for Seaton Delaval Coal Company, in Seaton Delaval House; Robert Foster Spence, agent for Backworth Colliery Ltd, in Backworth Lodge; and Edmund Octavius Southern, agent for Ashington Coal Company, in North Seaton Hall. Cramlington Hall and Arcot Hall were now occupied by a surgeon and a JP. The Samples were still residents of Bothal Castle and F.J. Snowball, agent for the Duke of Northumberland and Lord Hastings, had moved into Seaton Burn House.[6] While in many respects they were divorced from the daily life of the mining villages, as will be seen later, they played a critical role in constructing the boundaries within which the miners constructed their lives.

In fact, the actual physical construction of the mining villages depended upon the leases signed between the land and colliery owners. The leases spelled out, in detail, the actual powers given to the company. In some cases, such as Bedlington Coal Company, the land and colliery owners were one and the same. However, in some cases the landowners maintained control over the development of the villages. For example, the Duke of Portland was very much involved in the development of "his" villages of Pegswood and Ashington/Hirst. In 1873, the colliery owners of Pegswood Colliery approached the Duke for permission to build a school. Permission was given on the proviso that the Duke maintained control of the nomination of the management committee. No school was built. In Ashington/Hirst his refusal to allow public houses meant there were no "pubs" until after his death in 1889. At the same time he gave permission and support to the building of several community facilities: a school in Ashington (1873); a Primitive Methodist Chapel (1876); a Cooperative Store (1882); a recreation ground (1886); a drill hall (1889); People's Park (1896); and land for the Aged Miners Homes (1897).[7] His decisions were vital to the creation of an infrastructure of amenities in the villages. He was not the only landowner to impact upon the development of the villages. Lord Hasting's decision to have "Only one grocer's shop, one butcher's shop, and one public house" in "his" village of Seaton Delaval forced the miners of Seaton Delaval to build their Cooperative Society (1863) in nearby Seaton Terrace about half a mile from the village.[8] The Watsons of North Seaton Hall placed even greater restrictions on the village of North Seaton Colliery by stipulating that "no shop should be built" in the village – the nearest shops were half a mile away. Conditions were even worse in Dinnington Colliery where "nothing but colliery buildings can be built" – no schools, no shops, no public houses.[9] Of course, not all royalty holders were as strict or as interested in the development of the villages and in several cases complete power was given to the colliery owners. There is little evidence that the majority of owners were at all interested in the physical amenities in their villages. Perhaps the most important fact is that the miners had absolutely no say in the construction of the villages and the amenities of daily living.

Of course, things did change over the course of the century. The changes, for the most part, were a consequence of external events and, in particular, some Parliamentary Acts. One of the most important was the Mines Act of 1872 which stipulated that every colliery employing over 30 men must engage "a duly trained and certified manager." It was the appointment of these managers that precipitated to the fore a group of men who dominated life in the colliery villages until the First World War. Not only were they responsible for the mine but also for the conditions of living in the mining villages. For example, Richard Redmayne, resident manager at Seaton Delaval Colliery from 1894 to 1902, stated that apart from the technical work of colliery management there were:

> The schools to look after (I had three separate schools), the miners' cottages to inspect and keep in repair (we possessed 947 cottages), the farms to supervise (we farmed 1,100 acres) and Parish Council meetings to attend. I sat on two

> Parish Councils. . . . There was also co-operation in the social-cum-educational life of the workmen as directed through the medium of the Workmen's Institute, lectures to be arranged, University Extension classes to be organized.[10]

The first of these qualified managers was John George Weeks.[11] Weeks was appointed head viewer and agent for Bedlington Coal Company Ltd. in September, 1872. This was the beginning of a career that lasted 44 years until his death on July 8, 1916. From his appointment he was responsible for the building of cottages for miners. More importantly, he was elected to the Local Board in 1876 and, thus, was involved in the physical development of Bedlington, the sanitary conditions in town, and the development of roads. In 1879 he was appointed a member of Morpeth Board of Guardians which was responsible for the poor and out of work. He remained active in this role until his death. For the next 38 years he was involved in local government becoming the first chairman of the newly formed Urban District Council in 1894. Additionally, as colliery manager he was chairman of the managers of the Colliery School. This entailed frequent visits of inspection to the schools. In 1894 Weeks was made a Justice of the Peace for the county. Subsequently he was appointed chairman of the petty divisional court and a member of the licensing committee. He was actively involved in the Anglican Church acting as vicar's warden for several years. Finally he gave a helping hand to many of the self-help institutions in the district. The actions of Weeks impacted directly on the daily lives of the miners through his role in building houses, local government, schools, churches and other institutions, to say nothing of his primary function as manager of the mine. He was not atypical – other managers played similar roles in their own communities.

Throughout the nineteenth century the landowners, their agents, the colliery owners and their managers played the central role in creating the physical boundaries within which miners and their families lived. At the same time changes did occur that gradually eroded the absolute power of these groups. By 1914, the miners had some say in determining the conditions in the villages. The most important changes were precipitated by a number of Parliamentary Acts affecting decision making over land usage. The first semblance of change occurred after the passage of the Local Government Act of 1858 that empowered districts and communities to establish Local Boards to oversee local government in their communities. For the most part these became vehicles for the middle class to gain some control from the landowners. Miners were not successful in getting representation until 1894. The history of Cramlington Local Board, formed on September 26, 1865, illustrates the basic reality of power in the mining villages – control by the coal company.[12] From 1865 until 1890, the nine-man board was chaired by E.A. Potter, the colliery owner. He was joined on the board by members of the administrative staff of the colliery, other local landowners, estate agents, farmers and one or two businessmen but no miners. The first real change in decision-making power resulted from the Local Government Act of 1888.[13] This Act created County Councils and removed some of the power of landowners to make unilateral decisions over land use. However, the new County Council simply reflected a change in the location of decision making but not in who made the decisions. For example, the Northumberland County Council continued to be dominated by landowners and their representatives. Far more important to the villages was the Local Government Act of 1894 which placed decision making in the hands of locally elected Urban District Councils (UDCs) or Rural District Councils.[14] Miners soon appeared on the UDCs that were formed in communities with a substantial commercial or business community. In most mining villages, strong opposition was provided by the land and colliery owners. On several occasions between 1894 and 1912, when they finally gained UDC status, the miner members of Seaton Delaval Parish Council attempted to get UDC status. On each occasion, Lord Hastings and the colliery managers vigorously opposed their application.[15] At the same time, miners were beginning to make their presence felt and had a vehicle for making their views known, and slowly changes did occur in the villages. The playing field had become a little less uneven.

One further question must be addressed: how did the mining villages compare with other working-class enclaves with respect to decision making over land usage and their involvement in creating the conditions within which they lived. Certainly in the large cities decision making over land usage was more diversified. At the same time there was even less concern over the conditions of working-class life than in the mining villages. Additionally the strong middle-class moves for sanitary and housing reforms that emanated from the larger cities in the 1870s did not materialize in the colliery villages until the 1890s when miners were elected to the UDCs.

The mining villages and their inhabitants: the example of Seaton Delaval[16]

It is apparent from the above that throughout the century miners and their families lived in conditions not of their own making. Living conditions and the amenities available for leisure were dependent upon the colliery owners and officials. This was critical because of the central importance of the "home" in Victorian ideology. Thus, we must turn our attention to the actual conditions in the mining villages. It is impossible to provide an in-depth analysis of all the villages so I have chosen to examine one village – Seaton Delaval, a "typical" mining village – relatively small population, spasmodic and slow growth and nearly totally devoted to coal mining.[17]

The village of Seaton Delaval was created, by the coal company, between 1837 and 1841. It comprised 360 cottages of varying quality. For a small group, special houses were built – the doctor's house, the Manse for the Presbyterian minister, the Station Master's house, two houses for the headmasters of the schools and Wheatbridge House for the colliery agent. The rest housed coal miners and their families.

What, then, was the quality of these accommodations? Very little changed during the course of the century. The "better" cottages consisted of three rooms – two

Plate 1.2 Cambois, a "typical" mining village.
Reprinted with permission from Martin, E. (1997) *Images of England: Bedlingtonshire*, Stroud: Tempus Publishing

Plate 1.3 Foreman's Row, Seaton Delaval. Built in 1839–40. Picture 1951. Astley Arms in background.
Reprinted with permission from Allan, T. (1990) *Seaton Delaval: A History in Photographs*, Morpeth: Northumberland County Library

downstairs and an attic accessed by a ladder – no inside water, no sanitation. The inhabitants of Foreman's Row and Wheatridge Terrace had allotments attached to their cottages. The three rooms accommodated families varying in size from two to nineteen. Different families faced different living conditions. Some sense of the differences can be gauged from the 109 cottages in Foreman's Row, among the "better" accommodations in the village. Until the 1870s they accommodated, on average, around 500 people – just over five to a three-room cottage. With the opening of Forster Pit and the subsequent influx of miners the number living on the row increased to 620, an average of nearly six to a cottage. However, conditions were not the same for all families. In 1891, 140 (23 per cent) lived in relative comfort with four or fewer to a cottage.[18] The majority, 377 (60 per cent) lived in crowded conditions with more than five to a cottage. For example, Edward Ramshaw, a 51-year-old miner, lived in two rooms and an attic with his wife, six children and two lodgers. Of course, for individual miners conditions changed over time. Again, John Roseby and his wife Ann moved from Wallsend to Seaton Delaval in the early 1840s.[19] In 1851, they lived in Foreman's Row with their five sons aged 12 to 25 and two daughters, 22 and 9 – conditions hardly conducive to "family" life. After a brief sojourn in Australia in 1858 and 1859, the Rosebys returned to Foreman's Row. However, their circumstances were considerably different with only six people living in the cottage. By 1871, John, still working in the mine at 65, lived with his wife in number 91. Ten years later they were living in retirement at 104 Foreman's Row with a lodger, Charles N. Tait. By 1891, towards the end of an unusually long life, John, now 85, lived with his son-in-law William Dixon. Thus, the last 20 years of John's life were lived, physically, in relative comfort. His experience was typical with circumstances changing from extreme overcrowding to comparative comfort. However, the majority, when at the prime of their lives, lived in crowded conditions, hardly conducive to living the family life envisaged by the Victorian middle classes. Privacy and space were at a premium. It was virtually impossible to "get away" from anyone. Yet despite the, to the contemporary eye, abysmal conditions of the mining villages, they were a far cry from the truly horrendous conditions, even in the early twentieth century, of the cellars of Liverpool, the rookeries of East London, the back-to-backs of Birmingham and the tenements of Newcastle.[20] Thus, in something of an irony, the mining villages were by no means the worst places to live.

Of course, the cottages only provided the physical conditions within which they lived. Just as important was access to food and water, arrangements for sanitation, and certain basic services. Especially for women access to food and water were critical. Water had to be carried on a daily basis from pipes located throughout the village – imagine the problems facing women with six men working down the pit, all of whom had to have a bath when they returned from the pit. With respect to food, unlike many villages the residents of Seaton Delaval were never dependent upon company stores. However, in the royalty agreement, Lord Hastings had ruled that only three shops were to be allowed in the village – a tailor, butcher, and grocer/draper. The inadequacy of these services can be gauged by comparing Seaton Delaval village to South Blyth, Cowpen and Newsham, population 6,639, just over twice that of Seaton Delaval, which boasted 98 small businesses![21] For the first 20 years the women of the village remained dependent upon the three stores for supplies. In 1863, the miners and, I suspect, their wives, created the first miner-owned and operated Cooperative Society. However, due to the conditions of the lease, the Cooperative Building had to be built in nearby Seaton Terrace. This meant that women had to walk up to a mile to get their provisions! By 1876, Seaton Delaval Cooperative Society boasted a membership of 840, drawn from Seaton Delaval, Seaton Terrace and East Holywell. It is apparent that by the 1890s the conditions of the lease no longer applied, since the village boasted a dressmaker, a newsagent (still in operation today), a fruiterer and a stationer. Gradually, throughout the century the availability of goods and services improved.

Of course, the major question is to what degree was Seaton Delaval typical of other mining villages? The available evidence provides a sobering view of the conditions within which thousands of miners and their families lived. For example, on the opening of the Howard Pit at Netherton, "A short row of wooden back-to-back cottages with one room and a garret, brick floor, no ceiling nor services, relying for domestic water which came out of the pit"[22] were built for the influx of miners. Later in the 1870s:

> Cess pools were provided at street ends with the common outfall via the colliery drain. . . . Household water came from the pit pump . . . it was brackish and had to be boiled for domestic uses as it was largely impure. The unmade up roads were a nuisance and housewives made a path across the clarts with household ashes.[23]

In 1875 the owners of Dudley Colliery provided privies and ashpits for the first time. The conditions at Seghill, after 1858, deteriorated until it was recognized as one of the worst in the coalfield. The conditions at Netherton, Dudley or Seghill were not atypical. Conditions were even worse in villages where housing was provided by local landowners, such as John Clayton of Chesters. At Sleekburn he built "A street of back to back hovels consisting of one room and a lean-to pantry on the ground floor and a garret reached by ladder placed behind one door. The downstairs room was ceilingless."[24] Conditions throughout the coalfield began to be alleviated in the 1870s as a result of pressures from the newly established Local Boards. However, it was the passage of the Local Government Act of 1894 that brought fundamental changes to the ways in which the villages approached the problems facing the mining communities. The new UDCs and Parish Councils that were created in 1895 created committees to address the most pressing problems facing the villages. In nearly every instance committees were formed to address sanitation, water, roads and footpaths. The improvement in sanitary conditions, water supply and the state of roads was a priority across the coalfield and, thus, slowly conditions improved. During the first decade of the new century Councils expanded their committee structures. After the passage of the Small Holdings and Allotments Act in 1907, Allotment Committees were formed in Earsdon (1909), Bedlington (1910), Cramlington (1911), and Blyth (1912).[25] Allotments began to appear throughout the villages. Thus, it was only when external pressure was added to the demands that even minor changes were made. However, it is important to recognize that there was tremendous variability throughout the coalfield. It certainly does appear that improvements in sanitary conditions and the supply of water came to the mining villages with the creation of the UDCs in 1894, this was substantially later than moves to improve conditions in the larger cities. The colliery owners were not at all interested in the well-being of their workers.

The foregoing has provided a brief overview of the forces which underlay the development of the mining villages of East Northumberland; the dominant role of the land and colliery owners and the nearly total power of the on-site colliery officials. At the same time, there were small, subtle indications that the miners were not totally silent in the process. What emerged was a picture of villages whose conditions, to the modern mind, were grossly inadequate. However, one must remember that generations of miners forged good, productive lives in these, to outside observers, marginal living conditions. The majority of these observers, with a degree of surprise, commented on the cleanliness and comfort of many of the houses. The different perspectives were illustrated in 1873 when a reporter for the *Newcastle Weekly Chronicle* stated of 46 mining villages that:

> There is a large population of workers living, with a few exceptions in wretched houses, unprovided with any of those sanitary arrangements, which nowadays are considered as being almost the first necessities of a healthy social life.[26]

At the same time, other observers made frequent comments attesting to the cleanliness and brightness of the houses. In July 1873, at the Ninth Annual Picnic and Gala of the Miners Mutual Confident Association held on the Town Moor, Newcastle, the 10,000 assembled miners and their families passed three resolutions. Resolution number two expressed appreciation for the *improvement* in housing achieved during the past year![27] This highlights a critical point. While to many outside observers conditions looked bleak, to the miners things were getting better. We must always attempt to see things from the perspective of those living in the villages. In fact, as will be seen later, the living conditions in the villages were a formative factor in the development of a rich leisure life. In particular, the overcrowding of the "home" forced *men* out of the cottages in search of some pleasure and enjoyment. Of course, there is one ingredient, central to the development of life and leisure, that has not, as yet, entered the picture – the people who inhabited the villages. Who were they? Where did they come from? How stable were the communities? It is questions such as these that must be answered because it was people rooted in their own cultural backgrounds who engaged in a vast array of leisure activities. Thus we must turn to the inhabitants of the mining villages.

Some sense of who lived in the villages can be gauged by examining seven villages within a three mile radius of Seaton Delaval. In 1891, the percentage of the population born in each village varied widely: West Holywell (16 per cent); New

Hartley (24 per cent); Seaton Delaval (34 per cent); East Holywell (36 per cent); Seghill (40 per cent); Bates Cottage (52 per cent); and Old Hartley (66 per cent). Thus, in terms of birthplace, there was no such thing as a "typical" mining village.[28] Additionally, the villages were characterized by continual movement of people in and out of the village.[29] The reasons for moving were varied, sometimes voluntary and sometimes involuntary – miners in search of greener pastures, demand for coal which led to the opening and closing of mines and to increased and decreased demand for miners, and strikes. Strikes, in particular, often led to the total reconstitution of a village. However, the communities gradually became more stable, a critical element in the development of leisure.

Thus, one of the basic realities underlying life in the villages, instability and movement, was still considerable but, by the end of the century, was slowly decreasing. For a more nuanced view of the inhabitants let us turn again to Seaton Delaval.[30] In 1891, Seaton Delaval was home to 1,987 people; 683 (34 per cent) had been born in Seaton Delaval and 547 (28 per cent) in 15 villages lying within three miles of the village. Of the remaining 757, 7 per cent (149) had been born in 36 other Northumbrian mining villages. A small group of 103 came from 61 Northumbrian agricultural villages. A small number were born in North Shields, Newcastle, Morpeth and Blyth. Of those born outside of Northumberland, the greatest number came from the mining villages of Durham. Others were born in 13 other English counties with the largest number coming from Norfolk – the location of the Melton Constable estate of Lord Hastings. There were, also, a significant number from Cornwall and Devon who had come to Cramlington in 1865 and moved from there. Finally others had been born in Sweden, the United States, Canada and even on a ship off the Cape Colony in South Africa. In sum, Seaton Delaval was home for people born in 118 different towns or villages.[31] More importantly, the village was in a constant state of flux with people coming and going on a regular basis.

Of course, movement was predicated on the demand for coal. This was highly variable and is illustrated graphically in the demand for miners. The number of miners living in Seaton Delaval from 1841 to 1891 varied widely – 1841 (250), 1851 (544), 1861 (457), 1871 (411), 1881 (430), 1891 (681).[32] These figures demonstrate that tremendous changes, and thus movement, occurred particularly in the 1840s and 1880s. It seems that the villages were highly unstable. However, these figures do mask a basic process that was taking place – increasing stability. Increasingly people were born in the village, lived their lives there and died there and it was this group that played a central role in the development of leisure. The growing stability of the village population was exhibited in various ways. For example, the residents of Foreman's Row became more stable. In 1851, only two families who had been residents on the row in 1841 were still living there. This figure increased every decade; 24 in 1861, 34 in 1871, 46 in 1881 and 52 in 1891. Additionally 29 of those living on the row in 1891 had lived there for more than 20 years. In 1861 only 1 per cent of the heads of household had been born in the village. By 1891 this figure had increased to 13 per cent. Also in 1861 the majority of the population (63 per cent) had resided in the village for less than four years. By 1891, this number had dropped to 29 per cent. Similarly, in 1861, only 2 per cent of the population had lived in the village for 20 years. This increased to 14 per cent by 1891. Perhaps the most important fact is that over 30 per cent had been born in villages lying within three miles of Seaton Delaval. These had important consequences for life in the village. The movement of families from Seghill, Holywell, Newsham, etc. established relationships that tied the communities together – family, chapel and leisure brought the 16 villages into ongoing relationships. Thus life in the village was closely intertwined with that of other villages. The 1870s and 1880s must be considered as pivotal in the history of Seaton Delaval. It was during this period that it gained a degree of stability and a sense of community developed.

The birthplaces of the inhabitants of Seaton Delaval cast doubt on the homogeneity of the mining villages. Even within the village there were significant differences between people. This extended to where they chose to live. By 1891, there were two clearly established "desirable" places to live – Foreman's Row and Wheatridge Terrace. This was where people born in the village congregated – 56 per cent of the people born in the village lived on these rows when only 38 per cent of the total population lived there. From another perspective, of the 57 heads of house born in the village, 33 lived on the two rows. Additionally fewer people moved out of the village from these rows than did those who lived in other parts of the village. Even more significant was the fact that of the people who had lived in Seaton Delaval for over 15 years, 80 per cent lived on the two rows!

The view of a homogenous community is further shattered when occupation is taken into consideration. In fact, there were clearly defined positions of responsibility and status within the village. At the top of the hierarchy was a small group of "outsiders", individuals born outside the district, who were not miners. These comprised the "elite" of the village – the administrative staff of the colliery, the doctor, the schoolteachers, the stationmaster, and the ministers of the cloth, a total of 10. This small group, plus men who were not associated with coal mining, represented only 5 per cent of the working males in Seaton Delaval and 15 villages within a three mile radius. Miners dominated the villages.

While the differences between this "elite" and the rest of the population were to be expected, the differences amongst the miners were something of a surprise. There were various gradations based on one's role in the mine. There was a basic differentiation between those working underground and those working above ground. The difference was reflected in both status and money, with the underground workers tending to earn more and have greater status than their above ground compatriots. Underground there was a clear differentiation between the hewers on the one hand and the putters, drivers, onsetters and others on the other. They were all subject to a group of overmen and deputies. However, miners were not the only inhabitants of the village. In 1891 while the majority of the working male population were miners (681) a smaller group of 108 (14 per cent) were employed in a variety of occupations – 23 worked on the railway, 19 on the colliery farm and 29 provided services to the community as grocers, drapers, shoemakers, butchers, tailors and barbers.[33]

Throughout this analysis one group has been noticeable by its absence – women. In 1891, not surprisingly, 523 were wives, widows, or had no occupation. There was, however, a group of 123 (19 per cent) who were employed in agriculture (10), as teachers (11), domestic servants (67), dressmakers (27), clerks (6) and as a grocer's assistant and a hosier. For a small number there were opportunities to move beyond the boundaries of the home. For the domestic servants it meant leaving the village to go into service at either Newcastle or Blyth. However, these limited options and freedoms only lasted for a brief time and ended with marriage – the fate of the majority of women. Married women were not allowed to work and thus were forced to remain within the confines of their homes. The "home" was the focus of their lives and in the home they frequently ruled supreme. However, it is important to emphasize that there was never any mistaking their place in society – they were inferior and subordinate to men. As Billy Wilson of Cramlington stated of the early twentieth century: "We considered girls two or three places below us, they were not in the same category."[34] In fact, if a young girl was unfortunate enough to be the only girl in a family of males she became "nothing but a slave at work, to help her mother."[35] The whole focus of a woman's life was the two or three rooms that comprised her "home". The daily round of getting up to make breakfast for the men frequently meant getting up at different times during the night; 2 a.m. for the first shift of hewers, 4 a.m. for the putters and onsetters. As James Oswell of Ashington stated of his mother, in the early 1900s, she only took off her clothes on Saturday night because the five men were working different shifts.[36] The rest of the day was devoted to baking, scrubbing the stone floors, if they had any, and cleaning the house, an arduous task because the majority of villages were enveloped in a cloud of coal dust. Interestingly, outside observers frequently commented on the sparkling cleanliness of the cottages. When the men came home from the pit sufficient water had to be provided for hot baths. Since there was no water piped into the houses this meant a number of trips to the water outlets located throughout the village. Additionally women had to wash pit stockings, clean boots, and go shopping on a daily basis. They had no free time and thus no leisure except for the brief period between ending school and getting married. Not only did women face the reality of daily life but they were also excluded from the central institutions of male socialization: the ubiquitous inns and, after 1899, the social clubs. Thus they were excluded from the vibrant culture of male leisure. It is at this juncture that we can recognize a significant difference between miners' wives and other working-class women. Many working-class women worked. Thus, family life and attitudes to the "home" must have been different. Somewhat ironically, the much maligned miners' wives made the "home" the centre of their lives – very different from many other working-class women.

Notwithstanding the tremendous variability in terms of the actual composition of the villages, there was one set of relationships that dominated life and influenced the whole social milieu of the village – the relationship between the colliery owners and on-site officials, on the one hand, and the miners, on the other. This relationship was not a simple one. In fact, for long periods the relationship was cordial and even friendly. There are numerous examples that indicate that the miners held the officials in high esteem and high regard. However, there was never any question as to the nature of the relationship – the miners were always subordinate to the owners/officials. The latent hostility implicit in this type of relationship spilled over into violence during strikes. Strikes were an ongoing, relatively regular feature of life and their effects were felt for prolonged periods of time, some establishing perceptions and relationships that lasted throughout the century. In fact, there were different types of strikes which impacted differently upon the villages. Strictly localized strikes only impacted on the local community and often did not have long-term consequences. Their effect was cushioned by support received from other collieries. The same cannot be said about coalfield-wide strikes or lock-outs. These had a cataclysmic impact upon the whole coalfield because all were affected. The effects of these strikes reverberated through the coalfield for many years. All the coalfield-wide strikes were made worse by the fact that they all occurred during periods of depression in the coal trade and thus declining income for the miners.[37]

It is important to emphasize the starkness of the conditions within which miners and their families worked and lived. Additionally most villages were in the lee of smouldering pit heaps with the ever-present smoke. From the station at Seaton Delaval one could see 20 collieries "With their chimneys belching forth black clouds of smoke."[38] In stark contrast to the surroundings was the tidiness and cleanliness of the cottages. Outside observers made frequent comments on the cleanliness of the cottages.[39] Added to this was a strong sense of community. The colliery and colliery village must be seen as a complex social unit. As the late Stephen Martin, the miner historian, stated: "Amongst the mining fraternity a reference to a 'colliery' means not only the mine itself with its essential surface buildings, but also the community attached to it."[40] Mine and village were inextricably connected. While conditions were appalling, miners and their families strove to create meaningful lives. Sport and leisure, particularly for men, played an important role in their lives.

How similar were the mining villages to other working-class enclaves? In fact, the differences were greater than the similarities. Occupational homogeneity was only found in the textile towns of Lancashire and Yorkshire. The position of women was different from many working-class families. Housing conditions, while horrendous by contemporary middle-class standards, were better than those experienced in the working-class ghettos of the large cities. Additionally miners, at least the hewers, were significantly better off than the majority of workers in terms of hours of work and earnings. At the same time, few working-class groups were constantly aware of the presence, on a daily basis, of their "superiors". In the mining villages the teachers, ministers and colliery officials were a constant reminder of the basic social relationships underlying village life. Very simply the mining villages were significantly different from other working-class enclaves.

Why has it been necessary to provide an overview of the people who lived in the villages and the nature of the relationships underlying village life? It serves to highlight a number of factors that impacted upon leisure activities and, in particular, the necessary conditions for leisure: time, space and money. What the foregoing has illustrated is that the villages were not homogeneous communities – the differences translated into a complex hierarchy of power that had a profound influence upon leisure. In the first place, the power of the colliery owners and on-site officials was pervasive – little could be done without their approval. However, it is important to emphasize that their power was never total and never uncontested – the miners remained fiercely independent. Additionally, the power of the owners to make unilateral decisions with respect to conditions in the villages was eroded during the course of the century. Second, there was a complex system of status and power amongst the miners themselves. In a very general sense the community leaders tended to have certain characteristics – they were drawn from hewers, deputies and overmen – the positions with most status in the mine. Third, they were men who made the village their home; in other words, they were the long-term residents. Fourth, those born in the villages of Northumberland appeared to have more status than the others.[41] It was no accident, therefore, that it was when a degree of stability began to appear in the 1870s and 1880s that the infrastructure of leisure began to appear. Finally, and most importantly, mining society was predominantly a male society – women were, by definition, inferior. All of these factors came into play in determining the necessary conditions for leisure; time, space and money.

Necessary conditions of leisure

One of the necessary conditions of leisure is free time. Therefore we must turn to the hours worked by the miners and the number of holidays and other periods of free time they enjoyed. It must be emphasized that there was tremendous variability in the hours worked within individual mines and between different mines. However, despite these problems it is possible to sketch some basic elements and changes that occurred during the course of the century. In the early 1800s boys worked between 16 and 18 hours a day. Over the century hours were slowly reduced, to 14 hours in 1810 and 12 in 1831. By 1842, boys were going down the pit at 4 a.m. and coming up at 4 p.m. – a far cry from the hours of 1800 but still horrendous by modern standards. The workday was further reduced to 10 hours in 1872. Finally in 1908 miners gained an eight-hour day. It appears that it was during the second half of the century that the miners' hours of work allowed them free time in the evening and thus time to participate in free-time activities. In fact, this masks the true situation. For one group of miners, the hewers, an eight-hour day was a reality from early in the century. The hewers, the men who cut coal from the coal-face, comprised between 35 and 40 per cent of the workforce. From early in the century an eight-hour shift was the standard work day. This was further strengthened in the 1830s and 1840s when the two-shift system was adopted. What this meant was longer hours for the onsetters, putters, drivers, trappers and banksmen who had to service two shifts of hewers. Thus, for a substantial body of miners free time was never a real problem.

While hours worked and the amount produced were the foundation of the pay brought home, in reality, just as important was the number of days worked.

Throughout the century the basic work and pay period was a fortnight. This consisted of eleven working days, a pay Saturday and two Sundays – pay Saturday was often, but not always, a free day. The actual days worked depended upon the demand for coal. When the demand was high miners worked between 286 and 312 days a year. However, a contemporary observer reported that it was rare for any miner to work more than 280 days a year.[42] This allowed for 52 Sundays and 33 "other" days – pay Saturdays, religious holidays and local feasts. However, throughout, the mining industry was confronted with widely fluctuating demands for coal. The owners reacted to a decreased demand by closing the pit or laying off miners. As a result there was tremendous uncertainty in terms of work and thus pay. The variability in working days was most clearly illustrated in the 1870s, when conditions fluctuated between the best and the worst.[43] In 1873, the most prosperous year in the nineteenth century, the collieries worked, on average, 268 days. Four years later, in 1877, at the depth of the depression they were working only 191 days. A degree of stability came to the coalfield in the early 1900s when the days lost varied between 32 and 46 a year, a far cry from the wide fluctuations of the 1870s when the days lost varied between 45 in 1872 and 122 in 1877. In fact, hours of work varied on a fortnightly basis. Additionally, conditions were not uniform across the coalfield. There were significant differences between the steam coal collieries and the household coal collieries. Throughout, the actual number of days worked per fortnight was highly variable and always uncertain. More importantly the days not worked were unpaid. This time was recognized for what it was – "idle time", time not wanted and unacceptable.

By mid-century most of the traditional holidays, such as the binding period and the lengthy Christmas holidays, had disappeared. What happened during the second half of the century, and especially after the formation of the Union, was the creation of a number of holidays that were generally recognized across the coalfield. By 1898, it was claimed that the miners had 87 holidays a year – 52 Sundays, 26 pay Saturdays, Christmas Day, New Year's Day, Good Friday, Easter Monday, Whit Monday, August Bank Holiday, the Miners' Picnic, Newcastle races and a local feast.[44] Thus, the accepted working year was 278 days, still much greater than a modern person working a five-day week with four weeks' holiday – a total of 240 working days.

What can be concluded about the free time available throughout the nineteenth century? Even in 1914 they worked long hours at physically demanding jobs, thus the improvements were relative. However, it is evident that significant changes took place, especially after the 1870s. There was a gradual stabilization and standardization of hours and days worked. Although there was still uncertainty they were not subject to the wide fluctuations of the 1870s and before. Additionally, a basic calendar of holidays recognized throughout the coalfield had been established. After 1872 and the gaining of the 10-hour day miners, as a group, had increasing amounts of free time. Evenings opened up for leisure activities and every other Saturday was a focal point for free-time activities. However, this oversimplifies the reality. What must be taken into account are the different attitudes to free time related to the fluctuating demand for coal, the difference between free and idle times, and the difference between the hewers and the rest of the workforce.

Interestingly, whether the miners wanted free time or not varied throughout the period. At times, they made the acquisition of free time a major priority, while at others they just as vehemently fought attempts to give them more "free" time. This vacillation was related to economic conditions; in good times they wanted more time off work while, in bad times, the focus of attention turned to increased financial returns. This duality is masked when one uses the major strikes as an indicator of their concerns, because the major strikes all took place during depressions and thus the demands focused on money. The other side of the coin was reflected in the habitual absenteeism of the hewers. Absenteeism was such an engrained tradition within the coalfield that even the owners accepted it. While absenteeism was, in part, related to the severe nature of the work, which made it virtually impossible for hewers to work every shift in a fortnight, it was also an active choice related to the desire for leisure time.[45] Between 1873 and 1906, approximately 8.86 per cent of possible shifts were lost through absenteeism, one shift a fortnight. However, the active process of choosing free time took place only when they had sufficient money. Thus, the demand for free time reflected the same uncertainty and variability as other aspects of mining life. This does not apply to the whole workforce, but only to the elite workers, the hewers. For the other 60 per cent of the workforce it is doubtful whether free time was ever a high priority.

Finally in considering free time, as noted previously, it is important to differentiate between free time and idle time, or more importantly time with money and time without. Idle time was unwanted and the focal point of concerted efforts to eliminate it. The very term, used by contemporaries, "idle time" was used by the miners as a pejorative term to denote time that was useless, unfulfilling, and unwanted. While free time was used to create meaning, idle time was recognized for what it was – draining, useless, and threatening to a productive life. The miner had plenty of idle time during the nineteenth century. Central to idle time was the fact that it meant time without money and, as we shall see, money was central to leisure time activities. Thus, we must turn to the earnings of the miners.

While hours worked, holidays and free time were foundational to leisure activities, just as important was the availability of money, for the majority of leisure activities were predicated on money. Again we are confronted with major problems in providing an accurate view of what they earned. The inadequacy of source material is compounded by the variability between collieries, the differences between household and steam coal collieries, and the different approaches of different coal owners. However, it is possible to trace, in very broad terms, the earnings of piecework hewers, the most highly paid workers, from 1827 to 1914.[46] This analysis is based on the average earnings per shift of piecework hewers. While a more accurate picture would need to take into account the number of shifts worked per year, and as we have seen these were highly variable, the view that emerges is so clear that it can be taken as a rough estimate of annual earnings over the period. The figures are based on data which uses 1886 as a point of reference. The earnings for 1886 were set as 100. Different views of earnings, and thus purchasing power, emerge from different ways of collating the data. Looking at the average earnings on a decadal basis a clear picture emerges of steadily increasing income, rising from 50.7 in the 1820s to 135 in the first decade of the twentieth century. In fact, except for the 1870s and 1880s, growth was relatively consistent. Of course, in terms of daily living the gradual increase on a decadal basis had very little impact. Far more important were the fortnightly and annual earnings – it was the actual pay on a regular basis that determined whether they had money for leisure activities or not. The annual figures reveal a totally different picture, one of massive instability and uncertainty. In only 14 years (16 per cent) between 1827 and 1914 were earnings the same as the year before. Forty-one years (47 per cent) experienced increases and 32 (37 per cent) decreases in earnings. There was no way that miners could predict their earnings from year to year or even from pay to pay.

From 1830 to 1911 the coalfield experienced 10 cycles of increasing earnings followed by a decrease. There was no consistency or pattern to the changes varying from a total of 19 between 1836 and 1843 to a staggering 152 between 1870 and 1879. However, only once, in 1888, were the earnings at the end of the cycle less than at the beginning. In other words, over time the miners were being paid more. More important to daily life were the annual changes. The averages per year within the cycles varied from 2.4 between 1836 and 1843 to 15.2 during the 1870s. What emerges from this brief analysis is the fact that miners and their families lived within an economic environment of great uncertainty. It was impossible to predict earnings from year to year and, incidentally, from shift to shift. This uncertainty was magnified after the strike of 1877 with the adoption of a sliding scale that attached earnings to the price of coal. Since this fluctuated on a regular basis, so did earnings. This was unsatisfactory to the miners and culminated in the strike of 1887 after which a Conciliation Board was established and given the power to set earnings every three months. This meant that earnings fluctuated every three months. Added to this three-month variability was the variability in shifts working in different mines. What emerges from all this is the fact that miners lived in ongoing and perpetual uncertainty; there was absolutely no way that miners could plan for the future, every fortnight was a venture into the unknown.

This overview of earnings serves to focus attention on what was one of the most important eras in the nineteenth century – the period 1870 to 1879. While the cyclical nature of their earnings must have been a recognized part of their lives, nothing could have prepared them for the staggering events of the period 1870 to 1873 – an increase in earnings of 76 per cent? Never before, despite fluctuating earnings, did the miners have so much "ready" money – a necessary condition for many of their leisure activities. This unexpected prosperity came at a time when other important events were taking place – the recognition of the Union in 1872, the beginning of changes in local government, changes in education resulting from Forster's Education Bill and, more germane to leisure, the acquisition of shorter working hours and days in 1872, the development of a number of permanent buildings that were used for leisure activities – chapels and schools, and the opening of the first commercial sport facilities in the villages. Leisure was precipitated into a new era and never returned to the pre-1870s form, despite the cataclysmic erosion of their earnings between 1874 and 1879 – a 76 per cent decrease in earnings. However, there is little doubt that towards the end of the century the purchasing power of the miners increased dramatically. From 1830 to 1888 (58 years) the miners' income experienced an overall increase of 40. In the ensuing 22 years, their incomes increased by 48. Good times had come to the mining districts.

It is important to place miners within the context of other working-class groups in Britain. While, in very general terms, segments of the working classes benefited from a decrease in the hours worked and a general improvement in incomes in the later part of the century, there is little doubt that the hewers, at least, were, in good times, better off than the majority of workers in Britain.[47] This was important for the development of leisure as free time and "ready" money were critical ingredients in the development of a thriving leisure culture. Thus miners were in a premier position to exploit the increased opportunities for leisure activities in the 1870s and then, even more so, in the 1890s.

There was one further necessary ingredient for the practice of leisure activities – space. For middle-class Victorians the "home" was the centre of leisure activities. The crowded conditions of the mining villages meant that for the majority of miners and their families the Victorian ideal of family life was an impossible dream. At the same time, as we have already noted, for women the "home" was at the centre of their lives. However, the congested condition of many cottages forced men, in particular, out of the "home" in search of leisure activities. In the pre-1870 period facilities were few and far between.[48] By 1868 there were 23 churches/chapels and eight permanent school buildings. In a limited number of villages the colliery owners had built seven Mechanics' Institutes and five or six reading rooms. However, there were two facilities that appealed to a broader spectrum of the inhabitants of the villages – allotments and inns. Unlike many of the other coalfields the Northumberland coalfield was renowned for having allotments attached to the cottages. More pervasive were the ubiquitous inns. These were the centres of social life, providing a haven of light, warmth and conviviality in what can, charitably, be called a drab environment. By 1858, there were 78 inns, significantly more than the number of churches and chapels, spread throughout the length and breadth of the coalfield. Few villages did not contain at least one inn and no miner lived further than one mile from a friendly inn. In fact, inns were amongst the first buildings to be built after the sinking of the pit. They were the most important social institutions in the villages.

The late 1860s and early 1870s witnessed a massive expansion in the number of facilities for leisure activities. These new facilities precipitated the mining communities into what can only be called a leisure revolution.[49] There was a significant increase in the number of permanent church, chapel and school buildings. New Mechanics' Institutes sprang up across the coalfield and the first commercial sporting ground was opened. The period 1868 to 1875 witnessed an unprecedented building boom, in particular the building of 30 permanent churches and chapels, many with schoolrooms.[50] At the same time 35 new permanent school buildings, accounting for 63 per cent of all schools built during the century, were opened.

Plate 1.4 Seaton Delaval Miners' Institute. Built in 1900.
Reprinted with permission from Allan, T. (1990) *Seaton Delaval: A History in Photographs*, Morpeth: Northumberland County Library

The church halls and schoolrooms became the focal point of a rich leisure life albeit within clearly articulated ideological boundaries. They all fell within the compass of "Rational Recreation".

The early 1870s also witnessed the opening of seven Mechanics' Institutes. During the ensuing 20 years another 15 new Institutes were opened, all with the support of the colliery owners and officials. In all instances the Mechanics' Institutes suffered from too close an association with the owners and officials and thus were never popular with the majority of inhabitants in the villages. However, this total dependence upon the owners was gradually eroded with the miners taking a greater role in the building of the facilities. For example, in 1903, Viscount Ridley supplied the land for a new Institute at Cambois; building materials were provided by Cowpen Coal Company and financial support by Cambois Cooperative Society.[51] This sharing of costs was even more evident at Preston Colliery where three-quarters of the cost was provided by a loan from Northumberland Miners' Association. The rest came from the miners.[52]

Prior to the 1890s, the Mechanics' Institutes had one function – to provide educational opportunities. There was no place for frivolity. During the late 1880s and early 1890s a new dimension was added to the offerings of the Institutes – opportunities for amusement and entertainment. In a concrete sense this change was reflected most graphically in the introduction of billiards. *Every* institute opened after 1890 included at least one billiard table. While the inclusion of billiard tables led to an increase in membership in the Mechanics' and Miners' Institutes, their perceived relationship with the owners and officials and the central focus on educational activities meant that they did not appeal to the majority of miners. It was a change to the Friendly Societies Act in the late 1890s that precipitated the development of independent clubs organized, financed and controlled by miners – truly a revolution in facilities for recreation and amusement.

Annitsford Social Club, formed in 1898, was the first to take advantage of the changes in the Friendly Societies Act that allowed for the creation of independent social clubs. In the space of 18 years the institutional map of East Northumberland was changed in significant ways. The latent demand for this type of institution was reflected in the phenomenal growth in Ashington-Hirst where 18 social clubs were opened in the space of 12 years. By 1914 there were 43 clubs spread throughout the mining villages.[53] These clubs, totally independent of mine owners, colliery officials and publicans, became immensely popular, attracting over 50 per cent of the male population in many villages. At last, the miners had access to and control of space. These clubs were initiated, financed and managed by miners. They used organizational and administrative skills honed in running unions and cooperative societies to good effect. The characteristics are illustrated in the formation of two of the earliest clubs. In 1901, a group of Barrington and Choppington miners, under the leadership of William Temple, a hewer, purchased the German House at Summerside, Choppington Station, to use as a social club. The £1,500 was soon raised by issuing 1,500 one pound shares. All the shareholders were working men, the majority miners. In fact, the two largest shareholders, 45 and 25 shares, were two Barrington miners. The club was run by a Board of Directors which consisted of six miners and one mechanic.[54] Ashington and District Working Men's Club Ltd. was incorporated on January 18, 1902. The capital of 1,000 pounds was raised through issuing 1,000 shares. Unlike Barrington no shareholder could purchase more than one share. An entrance fee of 2/6d was charged. The Board of Directors comprised four miners, two engine-men, and one deputy overman.[55] The facilities provided at the clubs were remarkably similar and were illustrated in the new premises of Dudley and District Working Men's Social Club in 1909.[56] They included a billiard room with two large tables, a games room, three sitting rooms, a club room, a committee room, and a large quoit ground at the rear of the premises.

These were revolutionary changes and nowhere is the extent of these changes demonstrated more clearly than in the differences in the finances of the old institutions, the Mechanics' Institutes, and the new, the social clubs.[57] While Mechanics' Institutes were financially stable the money flow was not great. For example, in 1911, the Priestman Institute in Ashington, one of the largest and most successful in the county with a membership of 1,552 reported an annual income of £669.[58] A year earlier, in 1910, Seaton Terrace Social Club, a relatively small club by comparison, reported an annual income of £4,601 – significantly larger.[59] This difference reflected the reality; the social clubs were far more popular. At the same time it is important to emphasize that both types of clubs had one thing in common – they were exclusively male preserves. Women rarely entered these bastions of male privilege.

In the development of working-class controlled social clubs the miners of East Northumberland were simply doing what other working class groups were doing throughout Britain. The development of independent working-class clubs flowed from Henry Solly's social club movement of the 1870s which, by the 1890s, had been taken over by the working class. This is tremendously significant in that it is an indicator of how widespread the power of working-class inspired movements was. The last decade of the century was a watershed in the development of working-class action.

The development of social clubs was not the only change in terms of facilities available for recreational purposes. In late 1885, the Seaton Delaval Cooperative Society opened the Coop Hall in Seaton Terrace. The change was instantaneous, it became the site for an array of leisure activities. It was, in many respects, different from anything that went before – owned and operated by miners, it had a strong commercial orientation – in other words, it had to be financially viable. This was followed, in 1896, by the opening, by the Ashington Miners' Committee, of Miners' Hall in Ashington. By the turn of the century, two Coop Halls had been opened in Bedlington and Bedlington Station. While these initiatives were semi-commercial there were others that opened as commercial ventures. The real expansion of commercial activities occurred in the 1900s. The history of Ashington illustrates the massive changes that took place. The first commercial billiard saloon was opened in 1902. However, it was not until 1909 that commercial facilities for entertainment and amusement exploded onto the scene. This was the opening of the Princess Ballroom which was followed, in 1910, by the opening

Plate 1.5 Pavilion Theatre, Hirst, 1913.
Reprinted with permission from Kirkup, M. *The Biggest Mining Village in the World*, Newcastle, 1993

Plate 1.6 Astley Arms, Seaton Delaval with James Dawson, the proprietor. Reprinted with permission from Allan, T. (1990) *Seaton Delaval: A History in Photographs*, Morpeth: Northumberland County Library

of the Pavilion Theatre with seating for 1,400. This was designed with a dual purpose in mind – showing the new silent movies and hosting the theatrical groups that passed through the village on a regular though infrequent basis. By 1914, Ashington was served by five theatres and picture halls. Commercial entertainment had truly arrived, the whole focus of leisure had changed – the new world had arrived.[60]

Of course, one institution has been missing from the foregoing discussion – the inns. There is no doubt that they were the primary focus of leisure throughout the nineteenth century. But over time their premium place was threatened by the growth of alternative commercial facilities and, more importantly, the opening of the social clubs. However, even in 1914, they played important roles in the provision of facilities for meetings, dances, celebrations and sport. The 78 inns of 1858 increased to 128 by 1887 and 141 by 1902.[61] The majority of villages boasted one or two inns. However, there were some villages devoid of inns. Even in these instances miners were rarely far from a friendly pub – Bates Cottages was a short quarter of a mile from two inns in Holywell. East Cramlington Colliery was served by the nearby Astley Arms. In fact, no miner was more than one mile from a friendly hostelry. Like the Methodist Chapels the inns provided a focal point for the community, an ever-present symbol of stability in an uncertain world. However, unlike many other facilities, once the inn had been built and opened they rarely closed or changed function. Inns were an even more permanent fixture than the Methodist chapels. Even today, when some of the chapels and churches have been converted to other uses and the pit shafts have disappeared, the inns remain at the heart of social life in East Northumberland. In fact, the majority of inns built during the nineteenth century are still open today, albeit serving a different clientele.

The inns provide a bridge between non-sporting leisure and sport. Throughout the century they provided a meeting place for a variety of activities and also facilities for sporting competition. Although it is probable that they provided facilities from early in the century it was not until the 1860s that there was clear evidence of inns sponsoring contests on land adjacent to the inn. During the ensuing 50 years inns became the primary venue for sporting contests. For the most part the competitions required little more than an area of spare land. Since many of the inns were located adjacent to fields they were ideally suited to host the myriad of competitions that became popular after 1870. Nearly every inn hosted competitions sometime during the century but the majority held them on an infrequent basis. However, during the 1870s a group of 24 inns emerged that could truly be called the "sporting" inns of the coalfield.[62] The nature of the facilities can be illustrated through the history of the Astley Arms, Seaton Delaval (still in existence today as the Keel Row). It was the arrival of James Dawson from Norfolk in 1876 that precipitated the Astley Arms into a new role in sport. At first quoits were played on grass adjacent to the inn. In 1884, as a result of increased spectator demand, Dawson constructed a new quoit ground with facilities for spectators. Two years later he constructed a handball court with seating for 700 spectators! This was used on a regular basis until the late 1890s when the court was converted into a covered quoit ground with seating for spectators. At the same time, regular sparrow and pigeon shooting contests were held in the fields adjacent to the inn.[63] The other 23 "sporting" inns provided permanent facilities on a more limited basis. The development of these facilities reflected wider changes in society, in particular, those that impacted upon the use of space. Prior to the 1870s miners played their sports on public highways, public land, public beaches, disused waggonways and on ground adjacent to the inns. During the ensuing 40 years there was increasing pressure placed upon land which gradually forced the miners within the confines of enclosed sporting grounds. By 1914, the area boasted at least 14 public parks or recreation grounds, 11 community recreation grounds, 10 commercial sporting grounds, 50 enclosed football grounds, between 70 and 80 quoit grounds, 11 private cricket grounds, 2 golf courses, a number of lawn tennis courts and 64 Mechanics' Institutes or social clubs with billiard tables.[64]

Throughout the century the location of choice for miners' sports was public land – the Town Moor, Newcastle, Newbiggin Moor, the beaches at Whitley Bay, Seaton Sluice, Blyth, Cambois and North Seaton, and public highways. While freedom of access was zealously guarded, periodically the authorities made efforts to remove the miners from these traditional playgrounds. However, it was not until the 1840s that any concerted efforts were made by the authorities to remove them from public land. In 1844, a series of prosecutions were launched in Morpeth against the practice of running pedestrian contests on the roads leading into Morpeth.[65] These efforts failed. In the 1850s a series of prosecutions in the Seghill and Backworth areas succeeded in removing potshare bowling from the highways onto the moors and beaches.[66] In the late 1860s, as commercial traffic on the roads leading into Blyth increased, another series of prosecutions were laid against pedestrians (runners) who used the highways for their races. This time the authorities were successful. Later, in the late 1880s, at the height of the bicycling craze there were a number of serious accidents caused by "scorchers" – bicyclists going too fast. This resulted in the County Council banning cycle racing on the roads of East Northumberland in 1891.[67] Whether they were totally successful is a moot point since ongoing prosecutions for football, cricket, and pitch and toss attest to the inability of the authorities to completely control the highways. However, there is little doubt that the true miners' playground was the Town Moor at Newcastle and, to a lesser extent, Newbiggin Moor. These remained the mecca for potshare bowling until 1914 despite three concerted efforts by the authorities to remove them.[68] The moors remained, for the most part, free from interference. Access to the moors was critical to the development of working-class controlled sport. In this they were significantly different from other working-class groups. In most cities the commons had been enclosed or taken over for building. It was the exception rather than the rule for working-class groups to have access to public land free, for the most part, from outside interference.

One of the consequences of the increased pressure on public space and the banning of running on the highways was the development of enclosed commercial facilities. The first of these initiatives was undertaken by James Dowson, a Bedlington butcher, on May 17, 1873 when he opened the Bedlington West End Running Grounds.[69] This resulted from a series of events that led to an increased demand for facilities. It occurred during a period of unprecedented prosperity and the acquisition of the 11-day fortnight in 1872. The final ingredient was provided by Blyth police when they banned pedestrian contests on the public highways. During the next three years the West End Grounds were the focal point of fortnightly pedestrian contests. On May 30, 1874, James Holmes, a Bedlington publican, opened the East End Running Grounds at the other end of Bedlington.[70] This ground catered to the dog racing fraternity. Finally, in December 1875, three Killingworth publicans opened the Killingworth New Grounds for rabbit coursing and dog racing. These were the most successful grounds, hosting weekly events, which attracted large crowds from Newcastle who were brought to the grounds by special trains. By August, 1876 all three grounds were closed as a result of a concerted effort by authorities to close all commercial running grounds in the North East. The procedure used to close the grounds was simple and effective – charge the owners of the grounds with allowing betting, find them guilty, and close the grounds.[71] The success of the authorities was short-lived. By 1879, an innkeeper opened the Holystone Grounds near Backworth. There is little evidence to suggest that these early grounds were anything more than fields enclosed by a fence to allow for the charging of entrance money. On May 8, 1881, the commercial sporting centre of East Northumberland was opened.[72] On that day Joshua Rutter, the proprietor of the Bebside Inn, opened the Bebside Bicycle and Recreation Ground behind the Bebside Inn. The ground, with an eight lap to the mile bicycle track, became the only permanent fixture in the district. During the following

34 years, until its closure in 1915, a victim of war, the Bebside Ground was the focal point of all kinds of sporting competitions.

During the 1880s a small number of commercial grounds opened and then disappeared. These grounds were sensitive to the economic vagaries that swept through the coalfield on a regular basis. It was extremely difficult to launch commercial ventures with any chance of long-term success. The only ones to have success, apart from the Bebside Grounds, were five coursing grounds opened by innkeepers after 1897. The opening of these grounds reflected a process of centralization and the movement from public space to enclosed commercial facilities. By 1913, the five grounds hosted 91 per cent of all coursing competitions.

While entrepreneurs reacted to the increased demand for sport with the creation of commercial grounds, there were other initiatives that arose from within the communities themselves. There were two different initiatives; one, the creation of enclosed bicycle tracks met with short-term success and long-term failure. The other, community recreation grounds, was more successful in the long term. For a brief period, 1880 to 1895, bicycle racing became a major sport in some communities and led to the construction of 16 tracks. All the successful ventures were a result of cooperation between the different groups in the village. Without exception the construction of a track was dependent upon the Colliery owners providing access to land. The majority of tracks, 10, survived for only one or two years. The successful tracks were built by bicycle clubs with strong support from the owners. All, incidentally, were located in the heart of the coalfield.

More permanent and more important than the bicycle tracks were multi-purpose facilities created as a result of pressure from within the communities. Between 1886 and 1914, 11 community recreation grounds were created and experienced success. Again, their success was predicated upon cooperation between owners, officials and miners. On November 19, 1886, Ashington Recreation Grounds were officially opened. Thus commenced one of the most successful community enterprises. Central to its success was its financial viability. This was, in no small part, due to the continuing involvement of colliery officials and local tradesmen. Notable by their absence from any involvement on the committees that ran the grounds were miners. This first initiative was the only one that was totally within the jurisdiction of the colliery officials and tradespeople.

The other grounds were organized, financed and run by miners. However, in all cases, it was miners drawn from a particular group – those involved in the self-help institutions; the Mechanics' or Workmen's Institutes or the YMCA. The first of these, Bedlington Station Recreation Ground Company Ltd was formed in 1892 by a group of miners attending Bedlington YMCA.[73] During 1892, the company was formally incorporated and the capital of £300 raised by issuing one pound shares. Although 66 per cent of the shareholders were miners, they held only 25 per cent of the capital. The remaining 75 per cent was held by various officials at the colliery, the owners, and two spirit merchants. Like all community enterprises the Bedlington ground faced financial problems as a result of irregular use. It was the development of league football in the mid-1890s that brought financial stability to the company. The ground continued to be a sporting centre of the district until it was disbanded on March 29, 1916 due to the return of irregular use during the war.

Prior to 1894 and the passing of the Local Government Act that created UDCs there is no evidence that there was any interest in spending public money on recreational grounds.[74] By placing governance in the hands of locally elected officials the whole decision-making process over land usage was changed. One of the consequences of this change was the creation of three public parks and eight recreation grounds in five UDCs. Public parks with flower gardens, bandstands, lawn bowling greens and trees only developed under specific conditions – a relatively large population with a significant business and commercial community, for example Ashington (1896) and Blyth (1904). For the villages, ornamental parks were not an option. Their needs were more simple – recreation grounds with places for children to play. It was the plight of children that eventually moved the middle-class-dominated councils to reluctantly create recreation grounds for children. They were rarely created without strong vocal opposition. Recreation grounds were created in working-class Cowpen (1909), Newsham (1913) and Bebside (1914). Similarly, in 1909, after a lengthy campaign by the five miners on the council, Earsdon UDC opened grounds in Shiremoor, Holywell and Backworth. The difficulties faced by miner members of the council is illustrated in the case of Seaton Delaval Parish Council. From 1894 until 1912, when it gained UDC status, the council was chaired by Robert Embleton Ornsby, the resident agent of Seaton Delaval Coal Company. Throughout, miners were outnumbered by representatives of the coal company, local farmers and schoolteachers. The miners used the council as a forum to air their complaints about the coal company in an attempt to improve conditions in the village. Frequently, and surprisingly, the council concurred with the miners and sent delegations to the company in an attempt to improve conditions. Of course, both the chairman of the council and the representative of the company hearing the complaint were the same person, Robert Embleton Ornsby. As early as May 1, 1901, Thomas Chadwick, a member of the council and secretary of the local branch of the Northumberland Miners' Union, made an impassioned plea for a recreation ground for young and old. Recognizing the reality of the situation he suggested a piece of ground "which he thought would not be of very great profit to the Seaton Delaval Coal Company."[75] Council agreed unanimously and a delegation was sent to wait on Ornsby – nothing transpired. Nothing happened until 1912 when it gained UDC status. One of the first orders of business was the approval of a recreation ground. The ground was opened in 1914.

In most instances the facilities outlined above emerged as a response to specific needs articulated by specific groups within the communities. Commercial grounds resulted from entrepreneurs recognizing a need, bicycle grounds were related to specific short-term needs, community recreation grounds were more broadly based and rooted within the community, and public parks and recreation grounds developed as a result of the recognition of the need for recreational space. There was, however, a group of facilities that developed to meet a particular need and incorporated all the above groups. These were facilities for the "new" game of

Association Football that swept through the district in the 1880s and 1890s. In the 1880s games were played on friendly farmers' fields or on ground rented from colliery owners. As late as 1889, the premier club in Northumberland, Shankhouse Black Watch, played on the grounds of Stickley Farm. There were no recognized locations. With the explosion of interest in the game in the early 1890s and the emergence of leagues, permanent, enclosed facilities became a basic necessity for successful clubs. Gate money became a necessary ingredient for the continued success of clubs. This led to the development of permanent enclosed football grounds. Between 1893 and 1898 the number of these grounds increased from 3 to 30. They became the focal point of football in East Northumberland until 1914. The only other development was the opening of custom-made football stadiums: Portland Park, Ashington (1908) and Croft Park, Blyth (1909).

All of the above have, to a large extent, dealt with facilities used by miners. For a small segment of the population other facilities were created. The games of cricket, lawn tennis and golf appealed to particular segments of the population. Each entailed the building of specialized facilities and in each instance was given strong support by the land and colliery owners. Central to the development of cricket was access to an enclosed ground with a pavilion. A limited number of cricket grounds were developed. All had one thing in common – support of the social elite. In 1883, Sir M.W. Ridley provided the newly formed Blyth Cricket Club with a ground and a rent free lease. Later in 1912, the now Viscount Ridley, bore the cost of a new field for the same club. During the 1880s and 1890s the Duke of Portland, Lord Hastings, Sir M.W. Ridley and the Joiceys provided cricket pitches on the grounds of their estates. Golf invaded the northern reaches of the coalfield with the establishment, in the 1880s, of a links course on Newbiggin Moor. This remained the only course until the formation of Blyth Golf Club in 1905.[76] Finally lawn tennis courts came to the area in the late 1880s. At first they were limited to the grounds of Delaval Hall, Hartford Hall and the gardens of Anglican vicars in Bedlington and Sleekburn. The social elite provided for their own entertainment.

Conclusion

How can we encapsulate the foregoing discussion? In the first place, it is evident that miners and their families lived within a physical environment created by others. They had little say, until late in the century, over the conditions within which they lived. At the same time, because of the complete disinterest of owners and officials, they had freedom to live their own lives. While the conditions of living were abysmal, it must be remembered that to the miners they were the conditions they were familiar with. Remember that for the women these cottages, no matter how bad, were their "homes" and it is apparent that many created "homes". As to who these people were, they came from many different places and were continually on the move. As the century progressed the villages became more stable and there developed a core of people for whom the villages truly became home. These were, for the most part, native Northumbrians who, as will be seen, played a central role in the development of a vibrant leisure culture. With respect to time and money, the hewers always had free time and, in good times, a supply of "ready" money. Finally, in terms of space there were certain basic realities that had a profound impact upon leisure. Facilities, used for leisure, created during the century – chapels, schools, Mechanics' Institutes, bicycle tracks, football fields, etc. – *all* required the approval and active support of the dominant groups. Thus, many of the activities took place within the context of the values of the dominant hegemony. Even the inns, the central social institution, were owned and operated by "outsiders". However, towards the end of the century significant changes took place – the creation of facilities owned and operated by miners. Finally, throughout the century, those activities most deeply rooted in the communities took place on public land.

The question is to what degree were the mining communities similar to and different from other working-class enclaves.[77] Because of the lack of local studies it is impossible to make more than general comments. With respect to other colliery villages they were similar with respect to the relatively high earnings of hewers, the availability of free time, the shocking sanitary conditions, the centrality of the pub in the social life of the communities and the development of commercial facilities. However, there was no such thing as a typical mining village although the quality of housing in villages with rented housing, the majority, was worse than those tied houses in East Northumberland. Additionally, most mining villages were not comprised solely of miners. There were few villages where, as in the case of Seaton Delaval, 86 per cent of the male population was miners.[78] With respect to decision-making over land usage the clear patterns of decision-making in East Northumberland were not evident elsewhere. For the most part, "tied" housing was better than rented housing and in East Northumberland, often, the "tied" houses were rent free and came with a supply of coal. Additionally, with respect to shopping, the Coop Movement that dominated the Northumbrian villages was not as evident elsewhere.[79] In terms of the facilities for recreation, churches, chapels, and schools sprang up throughout the colliery districts of Britain, although whether they were used for recreational activities is unknown. By the early twentieth century most villages boasted football fields. However, it was only in limited locations that public parks or recreation grounds were provided. It is difficult to escape the conclusion that in nearly all respects the mining villages of East Northumberland were "better" than other mining villages.

It is when we come to compare the mining villages with other working-class enclaves that the picture becomes somewhat clearer. While each enclave was created within the context of particular circumstances, it is possible to discern, in general terms, similarities and differences between them and the mining villages. Even more than in the mining villages workers in the large towns lived within structural boundaries created by outsiders – a more diffuse and disinterested group of outsiders. Absentee landlords and landowners were the rule rather than the exception. Yet it was the appalling sanitary conditions of the urban ghettos that stimulated middle-class reformers to address the problems of the inner cities. Even though groups of people in Liverpool were still living in cellars and back-to-back housing was common in Birmingham and the West Riding in the first decade of the twentieth century, things were improving as a result of government action commencing in the 1870s.[80] The working classes of the large cities had access, from an earlier date, to a broader spectrum of public and commercial facilities. Public parks, baths and libraries were available in all cities by the 1870s, although it is debatable whether the working classes actually used these facilities. Commercial entertainment had been provided in London from early in the century; "penny gaffs" and street amusements were readily available. Population density was far greater in the urban centres and there were significant differences between urban workers and miners in terms of earnings and available free time. Miners in conjunction with engineers, some textile workers, and railwaymen earned more than other workers. In general terms miners lived in less dense and congested areas, a relative term and had, in good times, more money, and throughout the century had reasonable amounts of free time. The inhabitants of the cities, on the other hand, while living in congested accommodation were not subject, directly, to continual interaction with their "superiors". Additionally they had unfettered access to a large number of commercial amusements. It is important to emphasize the nature of the conditions within which they lived – abysmal. They literally forced people out of the home in search of life.

Was there anything that drew the working classes as a whole together? The answer is yes. However, perhaps the most important thing was the timing of events. New initiatives appeared first in Lancashire, Yorkshire and the Midlands – these were the focal points of change.[81] Across the spectrum of the working class slowly but surely workers earned more and gained more free time. At the same time, there was a downside: ongoing uncertainty in terms of earnings, hours of work, and even the availability of work itself. Thus, while over time things improved, in the ongoing reality of life they faced ongoing uncertainty. Second, throughout Britain municipal authorities were involved in the provision of recreational facilities often as a result of legislation passed by the government. At the centre of all these changes lay the middle class. Third, working-class social life focused on the pubs. In all working-class areas the pubs emerged as *the* central social institution.[82] As will be seen they were used by all groups as the focal point of a vast array of activities. In fact, they were more central in the larger urban areas where the ratio of pubs to people was as low as 1:90 while in the mining villages ratios were as high as 1:800. Perhaps more important was the development of working-class owned and operated social clubs. Although starting in London in the 1870s in middle-class dominated clubs, by the 1890s, they had been taken over by the workers. By 1914 they had spread to all working-class enclaves across the length and breadth of Britain. Similarly, while commercial entertainment had always been available, by 1914, in the form of cinemas, it had reached into the smallest communities.

Finally, while changes took place at different times in different places it was possible to pinpoint two particular periods when significant changes took place. In both cases they were related to economic conditions. Both in the mining communities and throughout England the period 1870 to 1873 was a period of unprecedented prosperity. The working class had never had it so good. This was paralleled by major changes in society, to such a degree that I refer to the period as a leisure revolution. As bad times came to the country in the late 1870s and early 1880s many of these changes were put on hold, but they never went back to the pre-1870 period. The next leisure explosion took place in the 1890s, a period of gradually improving times for the working class. While the evidence does not support a causal relationship the parallels are certainly suggestive. Good times bring significant changes.

2 Non-sporting leisure during the nineteenth century

Every miner has a hobby. Some are useful, some are not.[1]

The question is what were their hobbies, how extensive were they, and what role did they play in their lives? In fact, there were two, nearly separate, dimensions of leisure that reflected distinctly different views of the mining villages – non-sporting activities and sport.[2] There was a vaste array of non-sporting activities that emerged as soon as a village was built. These activities can be divided, roughly, into three areas: those focusing on self-improvement; those activities which had a direct impact on the quality of daily life, and those whose sole objective was to entertain and amuse. However, these activities were not ideologically neutral since the majority were practised within church and chapel halls, schoolrooms and, after 1885, the Cooperative Hall in Seaton Terrace.

Self-improvement[3]

From the early 1820s, the inhabitants of the villages were involved in education, temperance, and political activities. Central to all these initiatives was one group of men, Methodists. Their influence permeated all aspects of life.[4] They provided religious sustenance when the Anglican Church showed no interest, were at the forefront of the attempts to improve literacy, provided leaders of the Miners' Unions, were strong supporters of the Mechanics' Institutes, provided community leaders and were proponents of moves to increase literacy and provide educational opportunities. Thus, they were drawn from a particular segment of the community.

The drive for literacy and education was an ongoing concern, and forms one of the dominant threads that runs through the century. Any initiative aimed at providing education involved the colliery owners and managers; schools, reading rooms, libraries and Mechanics' Institutes were dependent on, and to a degree controlled by, the owners. Therefore the attitudes of the owners and managers were critical to any initiative. That education in any sense was not a high priority is not surprising. At the same time, for the colliery owners and officials, the problem of education was not as simple as it sounds. This ambivalence was reflected in a statement made by Nicholas Wood, Viewer of Killingworth Colliery in his statement to the Commission on Children's Employment in the Mines in 1842:

> Education would not benefit the men and boys as work people, nor would they be advanced simply for that education. Any higher kind of education would be of no service to them as pit-people; their work being mainly mechanical, and there being no superior mode of doing it. It is the opinion of many, and especially of underviewers, that they would be worse workmen if educated, as education would produce a dislike of work. But in all the popular commotions they have been worst in the most ignorant colliery districts. . . . In strikes the few men who may be educated may happen to be the most mischievous. . . . If any general system of education were deemed advisable for collieries there should be a general system of direction. Schools and libraries should be under the control of owners, and all blasphemous and immoral publications excluded. Many persons now come about with these things for sale, and they are purchased extensively.[5]

It is evident that the authorities were faced with a difficult dilemma. Their solution was that if there was to be any education it had to be firmly under the control of the colliery officials. This attitude permeated the nineteenth century – nearly all educational initiatives involved colliery owners and on-site managers in some way. It was within this context that a variety of educational opportunities were provided.

Efforts to provide educational opportunities typified all non-sporting leisure activities during the first half of the century – continuing efforts, short-term success and temporary failure. The experience of Seghill is typical.[6] In 1827, soon after the sinking of the mine, an evening school was opened in the village. This soon collapsed due to lack of interest. In 1834, the owners of the colliery opened a library. After a brief flurry of success it closed only to be reopened in 1837. It continued to run until 1843. It was re-opened in 1846. By 1850, the reading room was opened two nights a week and contained 800 books, newspapers and magazines. Seghill was not atypical. During the same period night schools were held in several villages but they too succumbed to the same malaise, lack of interest and low attendance. Comments were made about the lack of involvement of the miners, the majority of the users being drawn from the underviewers and administrative levels of the colliery. For example, in Seaton Delaval (1846) only 4 per cent of the miners enrolled for the lending library.[7] A library in Seaton Delaval in the 1850s did not prosper because "the men were suspicious of the owners' generosity." The attitude of the owners was clear; anything not under their control was suspect. At the same time there is evidence that miners were interested in education but only that free from control by the owners. The battle lines were demonstrated in a colliery near Seaton Delaval in the 1850s. T. Wetherby, a union pioneer, and a friend, both readers, decided to read to non-readers at night. They both put up five shillings, a considerable sum in those days, to buy books. When the Master heard of their intentions the two miners were summoned to his office and told "The men know quite enough already; if they learned any more they would know too much." If they did not desist they would lose their jobs. Their resistance to owner-controlled libraries was recognised by the owners. Mr Carr of Seghill bemoaned the failure of the library in the 1840s:

> The use of the library fell off at least twelve months before the strike (of 1844), when the men's minds got filled with what they were told by their leaders. They then read nothing but the Miners' Journal and the Miners' Advocate and the Chartist papers, such as the Northern Star.[8]

This suggests an interest among the miners in reading when the topics were of interest. In the 1840s one further effort, by the owners, to provide educational opportunities for the miners may be said to have fallen on fallow ground. Mechanics' Institutes were formed in nine villages. However, from the outset they were regarded with suspicion because they were perceived to be controlled by management.

The building of permanent school buildings in the period 1868 to 1875 greatly expanded the space available for recreational use. They soon became the centre of a broad spectrum of self-improvement activities: debating and literary societies, the University Extension Scheme, ambulance classes and, after 1890, offerings of the Technical Education Committee of the County Council. The first initiative, the University Extension Scheme, provides a classic illustration of the power relations within the mining districts and the conditions under which the managers, ministers and teachers got involved in the leisure activities of the miners.

The University Extension Scheme was started by Cambridge University in 1873. The movement reached the mining villages in 1880. In March 1880 a public meeting was held at Seaton Delaval Mechanics' Institute. Presided over by William Stoker, a miner and leading Primitive Methodist, it was moved and passed that:

> In the opinion of this meeting it is desirable that the University Scheme be extended into the colliery districts and that a committee be appointed with power to take such steps as they deem necessary to attain this object.[9]

Two miners travelled throughout the district promoting the scheme. Their efforts paid dividends in August when delegates from nine collieries met in the colliery schoolroom in Seaton Delaval to consider setting up a programme for the coming winter. The August meeting was very different from the March meeting. It was presided over not by a miner but by the local MP, who was supported by a professor from the College of Physical Science, Newcastle, five local ministers and one of the most powerful colliery owners. It was from this group that the officers to run the scheme were chosen. Support by the dominant groups from both within the community and outside was demonstrated vividly in 1884 by a donation of £50 from the colliery owners and a prize for the miner who stood highest in the class. The prize, donated by the Prime Minister, Gladstone, was a month's stay/study at Cambridge University.[10] This management domination was recognized by the miners in the continuing refusal of the Miners' Union to provide financial support to the scheme.[11] By late September 1880 centres had been established at five locations which provided relatively easy access to all miners in the coalfield. Until 1914 the basic structure of the University Extension Scheme remained the same, a series of between 8 and 11 lectures held during the period

January to April. The topics varied from year to year but were always provided by academics from Cambridge, Newcastle, London and Durham. Despite the fact that the scheme was plagued by low enrolments and poor attendance, it continued to provide intellectual sustenance to a small body of miners. It was through the University Extension Scheme that several miners embarked upon a summer of study at Cambridge University and thus for a small group it provided an avenue through which they could attain things they would never have attained. At the same time it demonstrates the gulf separating the managers and miners and the apparent lack of interest by large numbers of miners in things of an educational nature.

During the late 1880s, some new initiatives were undertaken: the development of mutual improvement and debating societies. The nature of these enterprises, their popularity and focus can be seen in the histories of the Bates Cottages Debating Society in the early 1890s and the initiatives taken in Seaton Delaval from 1896. The Bates Cottages Mutual Improvement Society was formed by Rev. G.A. Brown, the Anglican vicar of St Mary's Church in Bates Cottages, with strong support from J.B. Lattimer, the Managing Draper of the Coop Store in Seaton Terrace.[12] During the autumn of 1891 the society held fortnightly meetings. Early in 1892, the name was changed to the Bates Cottages Debating Society and the meeting time changed to every Tuesday evening. From January to April the membership of the society engaged in vigorous debates on a variety of topics and in a variety of forms. In the first week of February, Ralph Smith, a miner at Holywell Colliery, a prominent local preacher, reformer and advocate of cooperative principles read a strongly anti-church paper on "Disestablishment and Disendowment."[13] The debate raged for two weeks with the Rev. G.A. Brown entering the fray but all to no avail; the overwhelming majority voted in favour of disestablishment and disendowment. Certainly the miners of Bates Cottages did not toe the line. Later in February they simulated a House of Commons debate. On March 19, in the absence of a scheduled speaker, "Stump Orations" were introduced. This involved placing a number of topics in a hat and each individual drawing a topic and then speaking on that topic for five minutes. Again the range of topics suggests a lively debate: Music; Cooperation; Is smoking injurious?; Local option; What is your opinion of the teaching of atheists?; Astronomy; and Is marriage a failure? Some of the topics certainly have a contemporary twist to them.

Like many of the initiatives in the villages the Bates Cottages Debating Society disappeared from the scene in 1895, perhaps due to the departure of the Rev. G.A. Brown. A similar fate met the Seaton Delaval Literary and Debating Society which was formed in January 1896. The first debate, held in the colliery schoolroom, pitted the colliery manager, R.A.S. Redmayne against Frank Hartshorne, newsagent and teacher, on the topic "Was Hamlet mad?"[14] This was followed by a debate on "Total abstinence versus moderate drinking." Not surprisingly, considering the audience, total abstinence won hands down. The debating society, which started with such high expectations, soon collapsed due to its failure to attract an audience. This failure was revealed when the debate on "Women's Rights" failed to attract the expected numbers. Although, surprisingly, there was overwhelming approval to extend the franchise to women, the expected female audience did not materialize and the Blyth Weekly News in a most insightful comment stated that "for males, the education of the feet ie. football and dancing, is more attractive than education of the mind."[15] They were right, topics such as those addressed by the debating societies always appealed to a small group. However, there was always a small group that spent time engaged in intellectual exercises.

Perhaps even more interesting was the submerged and, for the most part invisible, world of miner-initiated and controlled self-improvement activities.[16] The autodidactic tradition was strong in the villages although it only comes into view through nearly accidental comments. For example, the prevalence of itinerant booksellers travelling through the villages on a regular basis is suggestive. In fact, the only real glimpse we get into the nature of these activities comes from two letters written in 1924 and 1925 by Harold Laski on his visits to Ashington. These letters provide a rare glimpse into the nature of these societies in the late nineteenth century. As he stated:

> I have had a dazzling time in the north, full of more genuine pleasure than I can remember in a long time. In Ashington, I stayed with a miner (Sam Snow) who began discussing Kant with me, with a knowledge and an insight which must be rare outside a group of dons. I gave four lectures to about 200 miners, and their questions and arguments were as searching as I ever met. Each evening I was there my host had in a dozen men for talk, and I frankly can rarely remember such talk. What was the value of Lester Ward's work? Where did Hardy stand in the pageant of English literature? Was security, as Bentham said, more important than freedom? Imagine a dozen men furiously smoking around a fire, and the questions beginning almost before I could think of the answers. And it was all done with the courtesy and modesty of really great gentlemen. Many of them had not had a day's schooling after fourteen years of age. They all belonged to a little society (Literary and Debating Society) where week by week they met to discuss intellectual problems. One man, the oldest of them, was a passionate Carlylian, and to hear him quoting rolling sentences from Chartism with the pleasant burr of the Northumbrian accent and his great gnarled hands making passionate gestures of approval was an experience I shall not easily forget. It reminded me all the time of that fine sentence of Carlyle: "Education is like light – from chaos it makes a world."
>
> I have just come back from my annual weekend with miners at Ashington – a really thrilling time. I gave them four lectures but learned more than I could ever teach. I stayed again with Sam Snow and his three thousand books, and talked with them each night until the dawn came. There was one to whom the truth had been brought by Lester Ward; another who had learned Greek in order to read Homer in the original; another who had the first edition of everything Hazlitt ever wrote and was prepared to back him against the world; another still, of Spanish descent, to whom life meant Don Quixote; and a grand

> Scottish Calvanist who thought that the whole truth was in Jonathan Edwards. It was sometimes grim talk, for there are hard times ahead for the mining community of England. But in general it was of books and men. Floyd tells the others that Nausicca makes other women futile – he being a bachelor; Snow says that Hazlitt on the English comic poets is the best dramatic criticism ever written; while McIntyre is prepared to prove the certainty of sin before the world. They are wonderful people. The dozen fellows I know best are all around sixty years of age, and like all men who live isolated and in the face of perpetual danger, they have really thought about the universe. These twelve, every Friday for thirty-six years have met to read and discuss a book. They argue grimly and you have to know your piece to get by them. They were saddened while I was there by the death of a miner who was found killed by a fall of stone; in his coat they found a translation of Thucydides with the page turned down at the Periclean speech.[17]

This reveals the hidden world of mining life. For a small number the pursuit of ideas was central to their whole lives. Imagine an academic lecture attracting an audience of 200! Imagine a library of 3,000 books in one small mining cottage! Imagine teaching oneself Greek and imagine taking a copy of Thucydides down into the pit! This provides a rare glimpse into the hidden world of life in the mining villages. Interestingly, unlike the activities promoted by the colliery officials, teachers and ministers whose efforts rarely lasted for more than a year or two, the Ashington society formed around 1890 was still thriving in 1925![18]

The colliery schoolrooms were the venue of another set of educational experiences that developed in the 1890s. These were different in their focus – they were pragmatically oriented providing the participants with practical skills rather than academic knowledge. Ambulance courses, the offerings of the Technical Education Committees and later the development of Nursing Associations provided avenues for the inhabitants of the mining villages to acquire particular skills that would transfer into daily life. In 1888, at the behest of the Seaton Delaval Coal Company, Dr Anderson, the local doctor, give five lectures in conjunction with the St John Ambulance Association to 80 miners employed at Seaton Delaval Colliery.[19] The coal company provided the equipment for the course and limited participation to their own workmen. By 1890, the work of the ambulance class was being celebrated by a public meeting where R.O. Lamb, the colliery owner, presented medallions and certificates to the successful candidates. The ambulance movement spread throughout the coalfield and on September 6, 1890 an "Ambulance Review" was held in a field near Bedlington Station with representatives from 10 collieries.[20] Like many other activities the growth of ambulance courses varied from village to village depending upon the attitude taken by the colliery officials. Although the work of teaching first aid was soon transferred to the Technical Education Committees the collieries continued to place an emphasis upon ambulance training. In May 1914, W.W. Burdon of Hartford House presented a silver challenge shield for competition between the collieries at Seaton Delaval, New Delaval and New Hartley.[21] The presentation was made in the colliery schoolroom.

While ambulance classes were directly related to safety in the mine, the Technical Education Scheme appealed to a wider audience and fulfilled a different function. Technical education was promoted by the County Council after its formation in 1888 and was organized and offered through the aegis of locally controlled Technical Education Committees. As a result of this local control programs varied from centre to centre. At the same time it was possible to identify some central elements. The variety of offerings and the basic characteristics can be gauged from the offerings of 1896. New Hartley offered drawing, dress cutting, ambulance classes, home nursing and cooking. New Delaval and Newsham, on the other hand, provided classes in ambulance, mining, shorthand, gardening and horticulture while Seaton Delaval offered machine construction and drawing, freehand model drawing and shading, girls/boys continuation class, and dress cutting.[22] These classes provided experiences for a wider segment of the village communities – women and men interested in "practical" affairs, and continuing education. At the same time technical education committees faced the now familiar problem of attracting people to their programmes and keeping interest when they were there. There were ongoing complaints over the problems of getting and maintaining students. In particular, it appears that women were not eager to grasp these new opportunities. Additionally they had problems getting teachers. These problems were reflected in the ever-changing offerings as the different committees strove to find courses that would appeal to a large enough audience.

The final group to use the colliery schoolrooms was significantly different from anything that had gone before, with the development of Nursing Associations. Rising out of perceived needs in the community this was one of the few activities outside of church and chapel to be dominated by women. In January 1902 the New Delaval and District Nursing Association was formed. Later in the year New Hartley formed a similar association.[23] In 1903 Mrs Jackson, the wife of the vicar of Delaval, was elected president of the newly formed Seaton Sluice and Old Hartley Nursing Association. Seghill Nursing Association was formed in 1906.[24] In each case the object of the organization was to provide nursing care in the community. The Nursing Associations had strong support from all segments of the community. Nurse Thompson, of New Delaval, was funded by local colliery owners, the miners' union and local business. In Seaton Sluice donations were voluntary but received from every householder in the community. New Hartley Nursing Association received its funds from the workmen at the colliery and other donations.

All these self-improvement activities, except for the independent debating societies, took place under the watchful eye of colliery officials, teachers and ministers, and were supported by relatively small numbers of miners. Interestingly, of the 21 miners/miners' wives, interviewed in the 1970s, only two (one became a deputy) participated in self-improvement activities,[25] although all commented, for the most part negatively, about their primary education. At the same time, there was evidence of interest in knowledge when under their own control. All the institutes created after 1900 boasted a reading room with a broad spectrum of newspapers and magazines selected by the membership – a far cry from the holdings of the Mechanics' Institutes where the colliery managers kept a close watch.

For a relatively small group, the eradication of the perceived plague of the villages, drinking, became the *raison d'être* of their free time. Working closely with the chapels and churches the temperance movement had a long history in the coalfield. In fact, it stretched back into the 1840s and 1850s. A meeting held in March 1871 demonstrated the basic characteristics of the movement: a permanent presence plagued by instability and fluctuating fortunes. Seventy members of the District Temperance Society met at Seaton Delaval to consider how best to rejuvenate the movement from its low ebb.[26] They first held a tea party in the colliery schoolroom before moving to the Primitive Methodist Chapel where they decided to form a temperance society with every religious body in the district. Like many other initiatives this failed to bear fruit. However, it demonstrated the conditions under which it did thrive, a symbiotic relationship between the churches and the temperance advocates. For a small but active group temperance became the *raison d'être* of their lives. The temperance movement took concrete shape in the 1870s. Sprinkled throughout the villages of Northumberland were Lodges of the International Order of Good Templars (IOGT) and the Independent Order of Rechabites. There were a small number of lodges that persisted from the 1880s – Newsham "Star of Hope" Lodge, New Hartley "Prospect Lodge" and Seaton Delaval "Reformer". These IOGT Lodges were joined by the New Hartley Rechabites.[27] In the 1880s many lodges initiated juvenile Templars and the churches and chapels formed Bands of Hope and branches of the Church of England Temperance Society. There were, in fact, two elements to the temperance movement. The first, and ostensibly the object of the movement, were the actions taken to eradicate drinking. These activities focused on public meetings and lectures aimed at converting the unconverted. Infrequently they joined forces in mass demonstrations and action against the drinking establishment. For example, in 1891, the temperance advocates of New Delaval – the Good Templars, Rechabites, Wesleyans, Primitives, Lay Churchites, Church of England, Methodist Free Church and Salvation Army – joined forces to fight against a new "off licence" in New Delaval. In this instance they succeeded.[28] Later in 1896, the temperance workers of New Hartley met under the chairmanship of Reverend G. Jackson, Vicar of Delaval, to discuss the closure of public houses on Sundays. This time they were not successful.[29] The second element of the temperance movement, and perhaps even more important, were the activities undertaken to provide social interaction with like-minded people. By the mid-1880s lodges were holding weekly meetings, fund-raising concerts, inter-club visits, and annual summer outings. How many people belonged to the lodges is difficult to determine but some idea can be given from a few brief insights. In 1884 a Grand Juvenile Templar demonstration and gala in Seaton Delaval attracted 500 children from Seaton Delaval, Shankhouse, Seghill, Hartley and Cramlington Templars.[30] The 1880 annual trip of the Seaton Delaval "Reformer Lodge" to Roker attracted 35 members.[31] In 1890 the Holywell Band of Hope claimed a membership of 80 men and 30 women.[32] The Northumberland District of the Independent Order of

Rechabites claimed 30 "tents" with 2,514 adult members and 1,798 juvenile members.[33] And in 1904 the 36 year old Bebside Sons of Temperance claimed 152 members. While these figures do not allow for a definitive analysis they are suggestive. While individual "tents" or "lodges" were relatively small the temperance movement did maintain a relatively strong presence in the district. They were never inconsequential and must have been a thorn in the side of the liquor interests.[34]

There was one other set of activities aimed at improving the conditions in the mining villages. These can, roughly, be described as political activities. In other words the ways in which miners attempted to influence the decisions that affected their lives both within the mine and in the villages. For a small but important group union activities lay at the heart of their lives, not just at the moments of strife and strikes, but in the day-to-day operations of union activities. These men were important in gradually changing the conditions of employment in the mine. However, more germane to this discussion were the efforts made by groups of miners to address pressing issues within the village. Prior to their election to School Boards, UDCs, and Parish Councils the only way they could make their concerns known was through public meetings. These meetings reflect the tip of the iceberg and were the culmination of weeks of debate and organization within the community. For example, in 1874 the inhabitants of Seghill met in the boys' schoolroom to petition for a School Board.[35] They were successful in their petition and a School Board was established, although not before having to go to court to force the colliery owner to allow the miners to vote for members on the Board. In a different context, 3,000 miners and their wives from Seaton Delaval, New Delaval and New Hartley, met at New Hartley to discuss the fact that some of the miners lived rent free while others had their rent paid by the employer and others, in the same colliery, Seaton Delaval Coal Company, were forced to pay rent.[36] About a mile away in Seaton Sluice, in 1886, the residents carried on a lengthy campaign to have "A good and sufficient supply of drinking water."[37] In this they were joined by the inhabitants of New Hartley, Seaton, Avenue Head, Seaton Terrace and Holywell. During the same week of March 1892 the residents of Seaton Delaval and Seaton Terrace met to consider different issues. In Seaton Terrace the concern was over the adequacy of school accommodation for Seaton Terrace children.[38] At present they had to travel four miles there and back to attend school. The inhabitants of Seaton Delaval, on the other hand, were concerned about the lack of space for meetings; the colliery schoolroom was insufficient for their needs. They "respectfully" begged the coal company to provide a suitable reading room and lecture hall.[39] This did not come to fruition until 1900! At this time these meetings were the only avenue available to the inhabitants to make their wishes known. This brief overview only provides a fleeting glimpse of the efforts of the inhabitants of the villages to improve their lives. Again it provides a view of a group of people deeply concerned about the conditions that influenced their lives.

The avenues to make their concerns known and to actually effect change opened up with the creation of School Boards and Urban District Councils. After the passing of the Local Government Act in 1894 miners began to appear on the UDCs and Parish Councils. The creation of Councils with decision-making power over the physical conditions within the villages placed the different groups within the village into a new arena, one in which they were forced to interact. This relationship and the time devoted to improving the conditions in which they lived were demonstrated in Seaton Delaval Parish Council from its inception in 1895 until 1908.[40] There were 9 of the 15 councillors who served for the whole period. There were three distinct groups amongst these long-term members. First, were the colliery officials; colliery manager and agent, the colliery engineer and the colliery cashier. The professional elements of village life were represented by two head-masters. They were joined by a local farmer. The miners were represented by an overman and two miners. Over the period these men met at the monthly council meetings and more frequently as members of the three committees – Sanitary, Roads and Water. Now the miners had a mechanism for the expression of their concerns and, just as importantly, they could not be ignored. For these men public service was central to their lives.

There were other political activities that attracted the attention of a small number of men and women. In 1878, a Conservative Association was formed for the districts of Backworth, Seghill and Earsdon. This was followed in 1886 by the formation of the Seaton Sluice Liberal Association, which attracted members from both Seaton Sluice and New Hartley. In November 1886 Charles Fenwick MP.[41] addressed the Association on rents and miners' earnings and the position of miners and owners generally. Perhaps more interesting was the formation of the Seaton Delaval Women's Liberal Association some time in the 1890s. They met regularly, on Thursday evenings, in the colliery schoolroom. The object of the association was related to local concerns – "to carry on warfare against the drink traffic, to secure the people better homes and better schools."[42] By 1910 they attracted up to 90 members to their weekly meetings – women were becoming involved in politics. It is likely that these formal organizations represented the tip of the iceberg with respect to political affairs. Certainly during the 1890s there were formal efforts by political parties to promote their own views. For example in July 1892, a Conservative "Van" travelled through the coalfield. On July 30 it visited Seaton Delaval. The speaker did not get an enthusiastic reception, being continually heckled by a large audience. As one heckler shouted "Is it right that foreign princes should come over and marry the Queen's daughter and then us poor devils have to pay £10,000 a year?"[43] Quite clearly the inhabitants of Seaton Delaval were not uninformed about the larger issues within the society.[44]

Gardening and friendly societies

There was another set of activities that can be traced back to the 1820s that had a practical focus: friendly societies and gardening. From the beginning of the development of the coalfield, miners created societies that provided for burial fees, injury benefits, and benefits for old age. While the primary function was protection and benefits these societies frequently became social clubs that met on a regular basis at the local inns. They were the first organizations to achieve a degree of permanence. For example, in 1839, one year after the sinking of the pit at Seaton Delaval, the Court of Squire Bates Ancient Order of Foresters was formed at the Astley Arms Inn. For the next 70 years the Society met at the Astley Arms on a regular basis and for generations of inhabitants provided a degree of security and support.

Similarly, gardening had a very practical and pragmatic function: to provide food for the table. The growing of vegetables and flowers and the keeping of livestock became a basic part of mining life. This interest was institutionalized in the development of flower shows. As early as 1825 comments were made about the emergence of flower shows in the villages. Interestingly, it was flowers, totally non-utilitarian, that were the centre piece of the flower shows, not the more utilitarian vegetables. From the earliest days, this reveals an element of mining life that usually lies hidden, their pursuit of things to brighten their lives. At first, the flower shows were plagued by instability but it was in the development of flower shows that the first signs of organizational permanence emerged. The first was the Seaton Burn Floral and Horticultural Society inaugurated in 1856. They were still holding shows in 1914. By the 1880s most of the villages had annual flower shows that became the focus of community spirit.

The nature of these flower shows, their role in community life and the changes that took place, can be illustrated through the history of the Seaton Delaval and New Hartley Floral and Horticultural Show which was held annually from 1876 on the second or third weekend in September.[45] The Seaton Delaval show was similar to all the other shows, the differences being of a minor nature. Thus it can be regarded as typical. The basic structure was put in place during the first few years. The show was held in a centrally located field, the use of which was granted by the colliery owners. The show's organizers erected two marquees for the exhibits. As the show expanded the number of marquees was increased to handle the exhibits. By the heyday, in the 1890s, up to five marquees were erected. At the heart of the show were the exhibits of flowers, fruits and vegetables. These were divided into various classes, which reached as many as 172 in the 1890s. However, other attractions soon developed, in particular a programme of athletic sports that soon began to attract competitors from all over the coalfield and beyond. In fact, by the turn of the century athletics replaced the flowers and vegetables as the focal point of the day. Other entertainment was provided by the Seaton Delaval Brass Band or some other band from the area. Finally, the show concluded, in the evening, with a ball in one of the marquees. The flower shows were the focal point of community life. For a small group of people, the exhibitors and the organizers, the show, itself, was the culmination of a year's work. For some growing flowers, fruits and vegetables was their central leisure time activity. And for an even smaller group the organization of the show was the focal point of their year's work, the members of the Seaton Delaval and New Hartley Floral and Horticultural Society. They met annually in January or February at the Hastings Arms to elect officers for the coming year and to plan the next show. This small group of 20 to 30 men undertook the organization of the show and the concerts that were held in the colliery schoolroom to raise funds. These became annual events. Additionally they frequently held celebratory dinners after the successful conclusion of the show in September. The form of the show was created during the early years. The 1891 show typifies these shows. There were three classes of exhibitors: cottagers, which was limited to residents of Seaton Delaval and New Hartley; and open, which attracted exhibitors from Dudley, Warkworth, Longhirst, Cowpen and Bedlington. Finally there was a farmers' class. The judges were brought in from Cowpen, Jarrow and Hebburn and were assisted by John Stephenson, the colliery cashier and the Rev. J. Brown, the Presbyterian minister. What did change was the nature of the programme – it was expanded to include different segments of the community. In 1884 a writing competition was held for the children of the two villages. This became a regular part of the programme, which over time expanded to include bread making, dressmaking and, by 1910, quilt making. The show appealed to all segments of the community – it truly was a community event.

The flower show was one of the few occasions when the community came together to celebrate itself. Up to 4,000 spectators attended the show – nearly the whole population of the two villages. It involved all elements of the community. The society and show were strongly supported by the colliery owners and managers, first through loaning the field and second through active involvement in the society itself. For example, in 1881 T.W. Asquith, a prominent landowner, was honorary president. On occasion they accepted official positions within the society. In 1894, R.A.S. Redmayne, the colliery manager, was president and John Stephenson, the colliery cashier acted as secretary. Additionally, the teachers, doctor and ministers played active roles. Duncan Robertson, the headmaster of the Presbyterian School was involved in an administrative role for over 20 years. Dr Anderson was a regular exhibitor. Duncan Robertson and Adam Galbraith, the headmaster of the New Hartley School, acted as judges of the writing competitions and the Rev. J. Brown acted as a judge of the exhibits. The non-miners of the villages were also actively involved. In 1891, George Forster, the village police constable entered exhibits in the open class. Railway workers, slaters, and other non-mining occupations also entered the competitions. The show was for the most part run by miners. But it did provide occasions where the different elements of the village could meet. G.T. Strother and Thomas Younger, two leading Primitive Methodists were actively involved in the society as too were some of the prominent potshare bowlers, quoiters, handball players and pedestrians whose lives focused on the inns. The show also provided meaningful roles for all ages, although exhibitors tended to be drawn from the older members of the community. In 1891, of the 33 entries in the cottagers' class over 70 per cent were over 40 and the average age was 47. This did not reflect the age distribution of the village. However, the younger members of Seaton Delaval found their outlet on the athletic field.

The history of the flower shows does not exhaust the interest in things horticultural. In fact, two other initiatives were undertaken to institutionalize the importance of gardens and gardening to the miners. Most important was the development of leek shows. Leek growing and showing had a long history in the coalfield. There was an exhibition of leeks at Dudley Colliery in 1858. Innkeepers in Choppington Station and Seghill held leek shows at their inns in 1867 and 1878. However, the real development of annual shows associated with leek clubs did not occur until the late 1880s. In 1889 both the Hastings Arms and Astley Arms hosted shows associated with clubs that met at their hostelries. Both the number of exhibitors and the prize money increased rapidly. In 1892 the 60 members of the Astley Leek Club vied for £30 in prize money. By 1896 the membership had increased to 103 and the prize money to £40. The number of leek clubs expanded rapidly especially after the opening of the workers' social clubs in the early 1900s. By 1913, between September 19 and October 17 there were 84 leek shows in the coalfield. Without exception the clubs were associated with inns or working men's social clubs.[46]

Entertainment and amusement

From the very beginning there is evidence of leisure activities designed solely to entertain and amuse. At first spasmodic and irregular, they eventually developed a degree of regularity and stability. Central to these activities were the inns and drinking. There is no doubt that the miners drank; the problem is in determining the extent and quantity of drinking. This is very difficult to determine, but in all probability less than contemporary middle-class observers thought. No matter how much they drank they had to be able to get up the next morning and go down the mine. There was one other activity that can be traced throughout the century – music.

Plate 2.1 Bigotphone Band, Seaton Delaval, *c.*1910. All the instruments, excluding the drums, were made of paper and card.
Reprinted with permission from Allan, T. (1990) *Seaton Delaval: A History in Photographs*, Morpeth: Northumberland County Library

Plate 2.2 Cowpen Colliery Brass Band, *c.*1900
Reprinted with permission from Balmer, R. (ed) *Blyth*, Zaltbommel, Netherlands, 1983

From the 1820s, it is clear that music was important in the lives of the miners. While it is difficult to trace, with any degree of accuracy, the existence of particular bands and choral societies, it is evident that they were an ever-present part of village life. Again we are confronted with the problem of evidence, enough to indicate that choral music and bands were an integral part of village life throughout the century, but not enough to get a clear picture of their role in society. However, some insights can be gauged from the history of Bedlington Brass Band between 1890 and 1899.[47] During this 10-year period 60 members joined the band. For many, 24, their tenure on the band was very brief, one year. In fact, the 19-player band was plagued by instability; only three members maintained membership in the band over the 10 years. However, there was a core of eight players who brought some stability. Band playing was a year-round pursuit and focused on the twice-a-week meetings or practices. In 1893, the band met to practise, have a meeting, or perform on 106 occasions. For example, in April they had eight practices and two performances. One of the major problems facing the band, and common to most activities in the village, was the availability of financial resources. Many of their activities entailed attempts to get money to buy instruments and pay the bandmaster. Income was raised from various sources; subscriptions from local breweries, merchants, publicans and the Bedlington Coal Company. Additionally fund-raising activities such as raffles and performing at flower shows, concerts and band festivals were sprinkled throughout the year. Also band members paid 6d a fortnight for the privilege of playing in the band. Christmas was one of the focal points of fund raising. At Christmas they paraded through the rows of Bedlington stopping at different locations to play a brief concert. The centrality of the inns and public houses to fund-raising activities and concerts was clear. In 1892 the band played at six Bedlington inns. Every year they played through the rows for the benefit of the public. Additionally, they played concerts or attended band competitions in such places as Hartford Hall, Blyth, Newbiggin, Amble and Cramlington Village. Their success was predicated on performance which was based on regular practices, held weekly on evenings, Baff Saturday and Sundays. Like so many other activities in the mining villages their success was, in part, dependent upon tacit acceptance by the colliery owners and managers. The band organizers went to great pains to ensure that they were in good standing with the colliery manager and owners. On several occasions they appeared at the residence of Mr Hare, the colliery manager, providing a concert solely for the enjoyment of the manager. The band also traveled to Hartford Hall, the home of "Squire" Burdon on the outskirts of Bedlington and even further to Hexham to the home of C.W.C. Henderson, the colliery owner, who was, unfortunately, not at home when they appeared. These efforts paid off when they wanted to get new members of the band. Frequently, deputations were sent to Mr Hare to see if he would provide a house and job for a prospective band member. On several occasions he met their request. It is obvious that for a small number of miners band playing was the *raison d'être* of their leisure. Additionally they played a small but important role in the life of the village appearing frequently to provide music on a variety of occasions. Bedlington was not atypical. For example, in 1893, Ashington boasted three bands: a colliery band, a string band and a temperance band. Members of the bands were provided with work and housing by the coal company.[48] Nearly every village could boast a band which performed the same role. In fact, bands, in all their forms, are one of the threads that runs through the century. They, too, reflected the competitive nature of miners. As early as 1859, four bands competed in the "Great Band Contest" at Cramlington Floral and Horticultural Show. In Ashington, in the early 1900s, band contests between the seven bands were eagerly awaited with the winner travelling to London to participate in national competitions.[49] Bands, like most other activities, revealed the competitive nature of miners.

Although the colliery schoolrooms continued to be used for educational efforts, from the mid-1880s leisure activities focusing on the schoolrooms expanded to include activities of a more "entertaining" nature and it was in the promotion of these activities that basic changes in ideology took place. The first signs of these changes came in the mid-1880s with the inception of Quadrille Assemblies.[50] The first, the New Delaval Quadrille Party was formed in 1886. By 1896, every village within three miles of Seaton Delaval boasted a Quadrille Assembly. While there were slight differences between the groups, they all held certain characteristics in common. The quadrille season usually lasted for about six months from the end of October to April and focused on weekly dances held in the schoolroom. The year culminated in an annual ball, which was one of the highlights of the winter season. Usually held on Friday evening the ball started at 9 in the evening. A break was taken for supper around midnight and then dancing resumed until either 4 or 6 o'clock in the morning. The ball was restricted to members of the Assembly or Party with the numbers varying from 20 to over 100. Additionally the programme of dances was carefully chosen and limited to about 26 dances, which included waltzes, the keel row and hornpipes – a varied programme. It became the practice to announce the Belle of the Ball: Miss Ida Lee at Seaton Delaval in 1894 and Miss Sarah Williamson at Bates Cottages in the same year.[51] Although the balls were private the public were allowed a limited role. In February 1894, 200 visitors witnessed the first three dances at Seaton Delaval Quadrille Party Ball. The Third Annual Victoria Assembly Ball, held in the Coop Hall, Seaton Terrace, was open to "a small privileged circle of spectators" – a clear recognition of the social differences within these small villages. By 1896, at the Eighth Annual Ball of the Seaton Delaval Quadrille Party, the general public paid to watch. By the mid-1880s the dancing on the green that typified many of the annual celebrations held in the mining villages had been formalized and institutionalized into regular patterns of behaviour. For a small group of villagers dancing had became a regular part of the winter season. In fact, dancing was more deeply engrained in mining culture and had always been a part of village celebrations. Its popularity is masked by the lack of evidence. However, it is evident that by the first decade of the twentieth century it had become an all-consuming passion for some members of the villages, in all probability those caught between school and marriage. For example, for James Oswell living in Ashington just prior to the First World War it was the most important leisure activity. He danced five nights a week![52] Similarly Billy Wilson of Cramlington stated that dances were a regular weekly feature in the Coop Hall.[53] Perhaps its pervasive popularity is illustrated most vividly in the opening of the Pavilion Ballroom in Ashington in 1909.[54] Here was a commercial operation devoted solely to dancing. As with so many activities dancing became regularized during the later part of the nineteenth century.

Starting in the 1880s there were other activities of a more entertaining nature that used the colliery schoolrooms: concerts and amateur theatricals. In February 1882 New Delaval Mechanics' Institute held a concert which was to include the Newcastle Portal Hand Bell Ringers. Unfortunately they were unable to appear because "their horse ran away near Gosforth".[55] In the same year, the Tomahawk Minstrels played to an overflowing crowd in the colliery schoolroom at Bates Cottages. Like many of these concerts they were held to raise funds for various village projects, in this case the new Workmen's Institute. The crowd was entertained with a sketch, "Massa's Golden Wedding." Later in February 1886, Newsham Black Diamond Amateur Minstrels held their inaugural performance in the schoolroom. The programme included John Turnbull, the champion clog dancer, "De saucy ginger blues" and Joseph Lacey's deliverance of a "stump oration" about "Women's rights."[56] Late in the 1880s amateur dramatics became very popular within the district. Cowpen Village Amateur Dramatic Society performed four plays in aid of Horton choir in February 1888.[57] This was the beginning of a number of amateur dramatic societies in several villages. Their repertoires varied from Sheridan's *School for Scandal* to temperance dramas entitled *Fast Life* and *April Fools*. Like many other activities the dramatic societies flourished for a brief time before disappearing and then being resurrected in a different form. Dramatic societies became a regular part of the leisure activities in the coalfield.

Although the importance of the colliery schools diminished with the development of alternative facilities they still played an important role in community life. By the early years of the twentieth century they were being used for public meetings, concerts, lectures and bazaars. Perhaps one of the most important functions was as a site for social functions sponsored by various clubs – homing societies, quadrille societies, five football clubs, a cricket club and a gymnastic club.[58] The leisure activities using the schoolrooms in 1914 could never have been envisaged in 1880; the whole world of leisure had changed.

However, it was the opening of the Coop Hall in 1885 that really revolutionized leisure in the district. The Coop Hall became the focal point for local activities not just for Seaton Terrace but for the surrounding villages as well.[59] Concerts were held to raise money for local causes: the aged people in 1886; Seghill Brass Band in 1894; and a benefit for a sick miner in 1896. The hall also became the location of annual events, for example, the St Patrick's Day celebrations for the Irish in East Northumberland – the National League of Cramlington (1888), Annitsford Branch of the Irish National League (1892), T.M. Healey Branch of the Irish National League (1894) and various other branches. It also hosted the annual show of the Victoria Cage Bird and Pigeon Show until the First World War. The hall was also used by various chapels when they needed larger accommodation. Both the Salvation Army and the Primitive Methodists used the hall. Public meetings were held to address such concerns as schooling, housing and sanitation. However, with the growth of alternative facilities around the turn of the century, the Coop Hall lost its premier position and simply became one of many venues utilized by the villages to host their leisure activities.

It was the Coop Hall that first attracted the professional entertainers who plied their trade in the villages of England. In January 1886, soon after the opening, Butterworth's "Selbino" Christy Minstrels entertained in the Hall.[60] They were followed, in March, by a thought reader who impressed the audience by finding a pin in Mrs Stewart's kitchen a quarter of a mile away. In the ensuing years the inhabitants of Seaton Delaval and vicinity were entertained by "Professor" Lenton the ventriloquist, conjurer, thought reader and clairvoyant; Carl Wright and his celebrated troupe of artists; Fred Archibold's company; and Fred Lindsay, comedian. In February 1892 the Hall hosted Paganini Redivivus "The Prince of violinists." These touring professional entertainers brought a welcome diversion from the harsh reality of life. In fact, it is possible to identify several groups and individuals who toured the mining villages over a period of years. "Professor" Lenton entertained on a regular basis throughout the villages of East Northumberland from 1886 to 1896. Fred Archibald's troupe were regular performers in the early 1890s. In June 1894, the Theatre Royal, a portable theatre operated by Mr Jones and Cassie Violet, opened for a season in the Coop Hall. Of course, not all professional entertainers focused on the Coop Hall, others set up in different locations. In Seghill, Tom Miller's Northumbrian Circus opened a short session in 1894. This "Circus" involved athletics, vocal turns, musical attractions and Harry Clinker, an Irish comedian. Seghill appeared to attract a particular group of entertainers, actors. Mr Hunter's travelling theatre appeared several times during the 1880s. In

1888 the troupe, comprised of Mr and Mrs Hunter and their three sons and one daughter stayed for seven weeks in Seghill. In 1891, Seghill was visited by a group of 13 actors, seven males and six females, all from outside Northumberland.[61] Seghill also attracted a series of speakers on "Spiritualism" in 1886. On several weekends during the summer J. Burns of London, E.H. Button of London, and Mrs Mary Mozart of the United States appeared in the village. Finally a number of circuses travelled through the district. In March 1890 a menagerie paid a one day visit to Seaton Delaval. There were upwards of 400 beasts, birds and reptiles including trained lions and educated elephants. In the following year, Sanger and Sons Circus, the "Largest in Europe" played in Seaton Delaval for one day. And in 1892, the Queen's Marionettes played in Seghill Market Place. What was claimed to be the best show ever to visit the village included a 42 stone (588 lbs), 19-year-old Madame de Albina, steam roundabouts, swings, a shooting gallery and a second-hand clothes stall. One can imagine the reaction of the natives of Seghill! The colliery villages were not isolated from the rest of England. They became part of the circuit of professional entertainers. In the larger villages, such as Ashington, professional entertainment was a regular occurrence. By the early 1900s the theatrical train used to come into town on Sunday bringing the actors, horses and other accoutrements. The most popular events were melodramas such as "Secrets of the Harem" and "The Stranglers of Paris." Buffalo Bill and his Wild West Show were a great success and Dr Bowing, the Great Hypnotist held the attention of the village for one week when he placed a girl in a trance and then exhibited her in the foyer of the theatre. Pitmen used to call in after their shift, at 1.30 in the morning, to check the spectacle. The offerings spanned the spectrum from melodramas to Buffalo Bill to Shaw and Ibsen, although this recital, at Hirst Coop Hall, had to be delayed because Miss Daisy Hallings had to "Plough through weary wastes of mud."[62] A revealing comment on the state of Ashington streets. The facilities for "professional" entertainment were expanded in the second decade of the century with the introduction of movie theatres. By the outbreak of the war they were springing up even in small hamlets like Seaton Terrace. Professional, commercial entertainment had reached into all corners of the coalfield.

Chapels and churches

Perhaps the changing nature of the society is illustrated most vividly in the changing attitude of the churches and chapels to leisure. By the turn of the century they were engaging in activities that 40 years earlier would have put them beyond the pale. From the earliest days the Primitives Methodists utilized religious services for social gatherings. Throughout the year various activities concluded with a tea and social. Prior to 1882 and the opening of the new chapel non-Sunday activities were restricted to socials and teas. It was in the 1880s that the activities became more secular. These changes focused on Saturday evening lectures and music concerts. Visiting lecturers talking on a variety of subjects began visiting the village on infrequent but regular occasions. While these lectures fell within the boundaries of "Rational Recreation," they were not exclusively religious in focus. As early as

1884 Thomas Burt, the mining leader and first miner to become an MP, attracted an "immense" audience to his talk on "Notes and Incidents on a visit to America."[63] Later in 1896, Robert Redmayne, the colliery manager and future inspector of mines, lectured on "South Africa."[64] These secular topics were intertwined with more clearly focused religious topics such as the Rev. Wainman of Blyth's speech on "The Devil: His Size, Weight, and Colour" in February 1888.[65] These lectures were an ongoing but infrequent part of chapel life that demonstrated that the villages were not as isolated as one suspected they would be. At the same time, the 1880s, concerts and magic lantern shows on non-religious topics were offered – the 60 voice Seaton Delaval Harmonic Society and Cramlington Philharmonic Society gave concerts. Also magic lantern shows on "Slum Life in London" and "Work of Life Boats" and "Newcastle Fire Brigade" expanded the horizons of the villagers.[66] From the 1880s the villagers were subjected to a steady flow of ideas brought by individuals from outside the community. Additionally the choirs became more active in the 1880s giving regular performances. The choirs travelled around other chapels and from the mid-1880s the eight choirs of the Seaton Delaval circuit were holding an annual picnic at Holywell Dene. By 1894 they were hosting a choral festival for choirs from Northumberland and Durham.[67] They were not isolated.

The balance between religion and secular activities is best illustrated in the Primitive Methodists' approach to children. The Sunday Schools were an important element of Methodist life with up to 70 per cent of children in the village attending. While clearly focused on inculcating religious beliefs, from as early as 1858, some "secular" activities were provided. In that year the Sunday School scholars were taken on a pleasure trip. The annual outing became institutionalized in the 1880s and became one of the highlights of the year. In August 1886, 200 teachers and scholars boarded the train for a trip to the seaside at Newbiggin. On arrival they marched in procession through the streets before participating in a variety of games.[68] By the mid-1890s, Tynemouth had become the venue of the annual trip. This secularization was reflected in the nearby Blyth circuit in 1895 when complaints were raised about the impact of football on Sunday School attendance. By the 1910s several Methodist Sunday Schools boasted football teams. There was one other element of Primitive Methodist activities that made them more visible in village life. From the earliest days, on special occasions, the Methodist Sunday Schools would parade through the village. One can imagine the interest in Hartley and Seaton Sluice in September 1880 when the three Sunday Schools celebrated the centenary of Sunday Schools with a parade of teachers and pupils with banners flying through the villages of Hartley and Seaton Sluice. Frequently, throughout the coalfield the Primitive Methodists marched through the villages bringing attention to themselves and their message.

Although the Presbyterians and Anglicans followed the Primitive Methodists in their emphasis upon lectures, concerts, and morally uplifting activities, it is clear that there were significant differences between them in their approach to leisure. This was partly due to the different orientations of their theology and partly to the social composition of their leading members. The Presbyterians numbered amongst their members the leading Scotsmen in the community; the medical doctor, the headmasters of the colliery school and the Presbyterian school. The Anglicans, on the other hand, included the colliery manager, engineer and cashier. In other words, the two churches appealed to different segments of the community. This difference was reflected in a more open approach to leisure activities. From the 1890s both hosted annual Garden Parties. The Presbyterians erected a marquee on the grounds of Hastings Cottage, the residence of Dr Anderson. Lawn games filled the afternoon and the evening concluded with a dance – a far cry from the amusements and entertainment of the Primitive Methodists. In fact, by the early 1900s, the Presbyterian Church boasted a literary and social club, a billiards table and, by 1911, Seaton Delaval Presbyterian Football Club.[69] The Anglicans, on the other hand, held activities on weeknights, not just on Saturday evening. By the turn of the century the parish magazine was publishing "Fixture Lists," lists of events to be held during the following month. In January 1902, for instance, the mission held a New Year's Social, a Sunday School treat and prize-giving, and a parish tea and entertainment. In February the parochial dance class finished for the season. On Saturday July 19, 1902 the St Stephen's Choir left by brake at 6.30 a.m. for their annual trip to Warkworth. They returned at 11 p.m. Two weeks later the Sunday School left by train for Jesmond Dene and even later the choir went on an excursion to Harrogate.[70] By 1913, the Monday night dance parties were going from strength to strength. There was an active men's society, a study circle, and a "Guild of Perseverance" (female). None of these could have been envisaged 30 years earlier.[71]

The opening of the Coop Hall in Seaton Terrace in 1885 had ushered in a new era with respect to the facilities available for leisure activities. During the ensuing 30 years a variety of facilities were either improved or built anew. In some cases these buildings were built by funds raised by miners, while in other cases the colliery owners built the improved facilities for their employees. During the 1890s, 15 Miners' Institutes and Mechanics' Institutes were built throughout the coalfield. Even more important were the social clubs that emerged after 1899. These were, without exception, built and run by miners. It was these social clubs that became the centres of social life for the men of the communities. In both the Mechanics' and Working Men's Institutes and the social clubs nightly programmes were offered – leisure became regularized. From being something that filled Friday evenings and Saturdays the leisure activities now spread throughout the week. These institutions offered local and national newspapers, a library, and a variety of recreational activities; chess, draughts, billiards, concerts, lectures, trips, inter-club competitions. In many instances they offered drink. The major difference from the Mechanics' Institutes is the fact that the clubs appealed to a far broader spectrum of the community. In fact, by 1914, it was estimated that over 50 per cent of the males were members of the clubs. They had revolutionized the leisure of the communities.

Community celebrations

It is apparent from the above that a leisure revolution took place in the 1880s and 1890s. In fact, leisure activities lay at the heart of significant changes within the communities, as is shown by the development of a number of annual community celebrations which emerged in the late 1870s. However, it is important not to overemphasize the break between the years pre-1870 and post-1870, because there were annual events in several towns and villages that can be traced back into earlier times and provide a thread of continuity between the earlier years and what transpired in the second half of the century. This is illustrated vividly by the history of the Bedlington hoppings which were held for three days every Whitsuntide from sometime in the seventeenth century.[72] The focal point of the three days of festivities was Front Street and the adjacent side streets. The hoppings attracted commercial attractions from outside Bedlington. The streets were filled with "numerous swing boats, galloping horses, shooting galleries, cocoa nut sallys, ice screamers, hokey-pokeyites, Jaffa orange vendors, ginger bread stalls."[73] Over the years menageries, circuses, theatres, boxing booths and a variety of other entertainments visited the hoppings. However, changes began to appear in the 1860s and it was in the athletic events that changes were to be observed. In the 1850s the programme consisted of a variety of footraces, three-legged races, old men's races, tilting the bucket and climbing the greasy pole for a leg of mutton. Over the next 50 years they became more "athletic" with the 120 yard handicap becoming the premier event. However, there were some things that did not change: the central role of the innkeepers and tradesmen in organizing and sponsoring the events. They were, from the outset, commercial enterprises. However, what is most significant is the real lack of change in the location. Despite efforts of the police in the 1890s to remove racing from Front Street and the various attempts to introduce alternative sports, the basic form of the hoppings remained unchanged. They provide a salutary lesson on the power of tradition in the mining communities.

The example of Bedlington demonstrates the dangers of generalizing to all mining villages while illustrating that some traditions persisted throughout the nineteenth century. However, the mining villages built after 1820 were forced to create their own annual calendar of events to provide a focus to the year. The only annual celebrations that remained from the old agricultural and pre-industrial calendar were Christmas, Easter, Whitsuntide, and Harvest. The saint days and traditional holidays had been obliterated. What emerged during the second half of the century was a calendar of annual events – annual trips, flower shows, children's days and aged people's treats.

The coming of the railway in 1847 expanded the horizon of the inhabitants of Seaton Delaval.[74] It literally opened up new vistas, which were soon eagerly explored by miners and their families. On one Saturday in the mid-1880s over 1,700 inhabitants of Seaton Delaval, more than half the population, bought tickets at the railway station, 700 for Blyth and the rest for Newcastle. The railways were soon utilized by groups in the villages for annual trips. As early as the 1860s the Primitive Methodists were taking the Sunday School children on annual trips. By

Plate 2.3 A Whitsuntide procession led by Netherton Colliery Band.
Reprinted with permission from Martin, E. (1997) *Images of England: Bedlingtonshire*, Stroud: Tempus Publishing

the 1880s nearly every organized group in Seaton Delaval and the surrounding villages went on trips – Sunday Schools, choirs, temperance lodges, Mechanics' Institutes, brass bands, cooperative employees, choral societies and the miners of Seaton Delaval Colliery. In some instances these became annual events that carried on for many years while in others they existed for a year or two before disappearing. The popularity of destinations such as Jesmond Dene, Roker, and Newbiggin suggest the embryonic development of a tourist industry to meet the needs of the visitors. During the late 1880s one of the highlights of the summer was the annual trip to Edinburgh of the Seaton Delaval miners and their families. In August 1890, 700 travelled from Seaton Delaval for an overnight stay in Edinburgh. By 1896 the scope of the trip had expanded to include Glasgow, Edinburgh, Stirling and Dundee.[75] Few of the inhabitants did not go on a trip during June and July and in many instances they must have gone on two or three trips.

In fact, the focal point of the summer was created even earlier, in 1864. Providing a symbolic focus for the miners of Northumberland was the Miners' Picnic held annually from 1864 on Blyth Links or Morpeth Common. The picnic became a centre piece of the whole year when up to 20,000 miners and their families met to celebrate their unity. The picnics began with the parade of separate collieries with their bands and banners flying. They marched from the station to the picnic grounds where the festivities were to be held. While the focal point of the day's festivities was the platform where leaders spoke on relevant issues, other activities

Plate 2.4 Children's Day, Bedlington.
Reprinted with permission from Martin, E. (1997) *Images of England: Bedlingtonshire*, Stroud: Tempus Publishing

were participated in on the periphery. Games, dancing, food stalls and other entertainments provided for the enjoyment of the large crowds. The Miners' Picnic was important in the calendars of the villages.

One of the most important changes occurred in the late 1880s, the celebration of Children's Day. Miners had always been concerned about their children and from the earliest years the Sunday Schools had provided trips, teas and games for their members. However, these opportunities were not provided for all children but only those attending particular schools. It was the passage of the Mundella Bill in 1880 that made education compulsory; thus, theoretically at least, all the young people in the village attended school. It was at this juncture that efforts were made to provide a special day for the children. From 1885, with the inception of the Holywell Children's Day, mining villages throughout Northumberland honoured their children with a special day. New Hartley (1888), Seaton Delaval (1889), Seaton Sluice (1889) and New Delaval (1891) followed the lead of Holywell.[76] The basic format was the same. For example in 1891 the children of Seaton Delaval paraded through the village to a field where they participated in sports, and Miss Davy achieved the remarkable feat of skipping 197 times in one minute. Later in June, in nearby New Hartley, the annual gala was held in the field which was used courtesy of J.H. Nicholson, manager of Seaton Delaval Coal Company. The children marched through the rows headed by the Wellesley Training Ship Band. In the field, the 500 children partook of a free tea. On the following Saturday the New Delaval Children's Treat Fund held the second annual gala in a field at New Delaval. About 500 children processed through the village headed by Blyth Artillery Band. They participated in boys' and girls' sports. Afterwards 450 prizes were presented to the children Not only did they celebrate community, but also their pride in being British – each child carried a small Union Jack. These were deliberate attempts to promote community spirit and nationalism. When they got to the designated field they were given teacakes. The whole proceedings were organized and funded by the community itself. The organization of the day was carried out through cooperation between different segments of the community. The colliery owners and managers provided financial support and facilities, a field. The actual organization and fund raising was carried out by a School Treat Committee. For example in 1890, the honorary officers of the Seaton Delaval Treat Committee were the agent, the colliery manager and the colliery cashier. The active members of the committee were four coal-miners. Basically the success of the Seaton Delaval Children's Day and those in other villages depended upon cooperation between all elements of the community. Fund raising was done in various ways: dances, concerts, and subscriptions. The method of raising funds varied from village to village. For example, in New Hartley, the money was raised through subscription and a levy paid by all miners working at the mine. In Seaton Delaval it was done through concerts. These days continued until the war, although like most things in the mining villages they experienced ups and downs. The success of these ventures depended upon the dedication of the School Treat Committee. Membership of the committee changed from year to year and thus the success varied from year to year. Not only did membership change but the actual organizations that sponsored Children's Day changed over time. In Ashington different groups assumed the leadership at different points in time. By the late 1890s, the day was a very small celebration and, in fact, Children's Day was not celebrated in 1903. The next year the local Council took the lead and sponsored a Children's Day. By 1909, a new Children's Sports Committee with representatives from the thirteen social clubs had been formed. Thus, the organization shifted from the Council to groups solidly rooted in the community. The Ashington–Hirst Children's Day Sports Committee, in 1913, comprised representatives from the social clubs and various societies.[77] They went out of their way to provide a day to remember. The day started with the schoolchildren meeting at their own schools. In Hirst the pupils and teachers marshalled at the six schools before marching in procession to People's Park where the sports were held. Each school was led by the headmaster, teachers and a band. The Ashington schools procession was led by the Harmonic Band and the Linton and Ellington Schools by the Salvation Army Band. One can imagine the excitement in the village as the various processions wended their ways to the park. At the park each child received 1d, a bun, and some sweets. The committee had raised £130 to provide 7,200 children with the treats. The parents turned out in full force for this gala celebration. There were races for all and even a baby show, which was judged by Drs Spence and Thompson. After the festivities the children and their parents spread out throughout the show grounds to listen to the organ that played all day, gawk at the fat lady, ride on the roundabouts, visit the coconut stall and shooting galleries and shop at the stalls set up throughout the park. The day had truly become a communal event bringing the whole community together to celebrate communal identity. Ashington was not atypical, Children's Galas were an important day on the calendars of most mining villages.

The final event that provided a focus for the celebration of community was different yet reflects the basic values underlying the mining villages. In February, March and April 1886 a series of three concerts, sponsored by the Cooperative Society, were held in the Coop Hall, Seaton Terrace to provide funds for a dinner for local widows, men over 60 who were unable to work and their wives. A group of 17 men and women from Seaton Delaval and Holywell sat down to a dinner on April 24, 1886 in the Coop Hall.[78] This was the first of many such efforts. The concern over the lives of older people gathered strength during the 1890s as the number of "aged" miners increased. It was in 1899 that the Old People's Treat was institutionalized in Seghill and an organization set up to gather funds and make arrangements. This was followed by the Seaton Delaval Aged People Treat Fund in 1906, the New Delaval and Newsham Aged Folks' Fund by 1910 and the Annual Old People's Day in New Hartley at the same time. Some villages provided summer outings and winter indoor treats but every fund or day provided an annual dinner – this was the highlight of the year. The 17 people, who sat down in 1886 at the Coop Hall, Seaton Terrace, were overshadowed by the 250 who attended the seventh Annual Treat in December 1912.[79] The treat was truly a sumptuous affair providing roast beef, mutton, pork, rabbit pies, vegetables, plum pudding, jellies, blancmange and fruit and mineral waters plus, for the men, cigars, tobacco and pipes. After the meal they were escorted to the newly opened Queen's Hall Picture Palace to witness the wonders of cinematography. They returned to the Miners' Hall for tea and gifts of chocolate, cigars and packets of tea – a truly memorable day for the older citizens of the village.

Inns

To anyone familiar with the mining villages, even today, there is one institution conspicuous by its absence in the above account – the inn or public house.[80] Lectures, concerts, educational schemes, temperance meetings and other elements of leisure life utilized the chapel and church halls, the schoolrooms and Mechanics' Institutes. All were ideologically opposed to drink. The inn provides a bridge between those activities supported by the dominant groups in society, a minority, and the activities of the majority. It was a catalyst in village life that bridged social and ideological boundaries. The inns were used by all segments of society – mine managers, medical doctors, schoolteachers, ministers, underviewers, overmen, hewers and all other workers in the mine. The Astley Arms (1838) and the Hastings Arms (1840s) were built by the coal company and became the focal point of community life throughout the century.[81]

Prior to the building of rooms attached to the chapels and the colliery school (1867), "The long room at the Hastings Arms was the only place" where meetings could be held.[82] The long rooms were utilized by all groups – for friendly society meetings, inquests, and, on occasion, church services. Interestingly the two inns served different functions. However, certain activities were common to the two inns, in particular social activities involving the miners as a group, activities with little outside influence. Throughout the century the miners formed friendly societies; the Court of Squire Bates, No. 958, Ancient Order of Foresters met at the Astley Arms on a regular basis from 1840 until the First World War. Lodges of both the Foresters and Oddfellows held regular meetings, half yearly district meetings, and annual suppers at various hostelries throughout the coalfield. Similarly, from the 1880s, lodges of the Loyal Orangemen and Irish National League held meetings at the inns. The Seaton Delaval Burns Club held annual celebrations of the bard's birth from its inception in 1870 until 1914. The two Seaton Delaval inns were also the location of a variety of annual dinners. At various times groups of workmen from Seaton Delaval Colliery held annual suppers; mechanics and engine drivers in 1878, locomotive men and mechanics, and locomotive men and fitters in 1910. As the community became more stable and it began to have a collective history the inhabitants began to celebrate long-term service. For example, in 1892, the Seaton Delaval Branch of the Northumberland Engineers' Association made a presentation at the Astley Arms to John Fairbridge of New Delaval for 40 years' service as a brakeman.[83]

The inns played a central role in the development of flower shows.[84] As early as 1862 the Hastings Arms hosted the annual show of the Seaton Delaval Floral and Horticultural Society. This organization died in the late 1860s. A more stable foundation was laid in 1876 with the formation of the Seaton Delaval and New Hartley Floral and Horticultural Society. The new innkeepers of the Astley Arms and Hastings Arms, James Dawson and William Harper, played leading roles in promoting the organization. In 1877 they, jointly, provided a silver cup for competition. At the same time, Dawson provided the field for the show and a marquee to hold the exhibits. From 1879 the more centrally located Hastings Arms became the centre of the organization by hosting the annual meetings. The innkeepers also played a central role in the burgeoning interest in leeks in the 1890s. The two inns became the focal point of leek clubs and shows. In fact, in 1897 Dawson erected a wooden building to host what was claimed to be "the largest leek show in the mining districts." Thus the inns were the centre for a wide variety of social activities.

Although both inns supported a wide variety of social activities, there were significant differences between them as noted. From the earliest days the Hastings Arms was associated with political activities. This may have been due to its more central position in the village. In any event over the years, at times of crisis, the miners met in the long room of the Hastings Arms.[85] They met in 1865 to discuss the actions to be taken with respect to the bitter Cramlington strike. During 1877, as the miners moved inexorably towards strike action, the Hastings Arms was the centre of political activity. During the strike Mr Harper, the innkeeper, was one of the strongest supporters of the Seaton Delaval and New Hartley soup kitchen. Later in the year a meeting was held to support the candidature of Albert Grey for parliament. Representatives of 26 collieries met, in 1896, to discuss ways of arousing interest in the Allotments and Small Holdings Act. In the following year a conference of representatives was held in the long room. As late as 1910, the Seaton Delaval Branch of the Wansbeck Unionist Association held a meeting at the inn. Interestingly no political activity was recorded at the Astley Arms.

The Astley Arms was different in two major ways. First it had a long-term relationship with a number of organizations and second, several of the organizations involved the upper echelons of Seaton Delaval society. From 1840 until 1914 the Court of Squire Gates No. 958, Ancient Order of Foresters held their regular and annual meetings at the Astley Arms. They were followed in 1867 by the annual supper of the Cramlington locomotive drivers and stokers. This was held annually until 1895. With the expansion of the colliery in the 1870s, a number of clubs supported by colliery officials, were regular patrons of the inn: the Burns Club (1869–1914); the Astley Social Club (1878–*c*.1900); and the Hartley Hastings Cricket Club (1881–97). The role of the upper echelons of the coal company in these endeavours is illustrated through the involvement of Thomas Beck, the colliery engineer. Beck was the driving force behind the formation and continued existence of the Astley Social Club. This club, formed in 1878, comprised eight gentlemen from Gateshead: James Dawson, the innkeeper, two local farmers, the foreman mason, an overman and several others. This brought together representatives from different constituencies within the community. The club met on a regular basis at the Astley Arms. Thomas Beck was also the chair of the 1891 annual supper of the Seaton Delaval Coal Company mechanics and locomotive engineers. He was still performing the role in 1902 at the age of 70. Beck was also the chair of the Hartley Hastings Cricket Club. In 1898, the Astley Arms was chosen as the location of presentations to two of its leading citizens – the local doctor, Dr Anderson, and Robert Embleton Ornsby, the agent for the company and the most influential person in Seaton Delaval.[86]

The central importance of the inn as a catalyst for the village was reflected most vividly in the importance of Mr and Mrs Dawson in the life of the community. Between 1876 and 1899, their contributions were recognized on several occasions. In 1878 the committee and friends of Seaton Delaval Colliery held a supper, at the inn, to present James Dawson with a gold medal and Mrs Dawson with a gold medal and silver set of tea and coffee service. This was presented by the workmen of the colliery in appreciation of their kindness during the strike of 1878. Later in 1880, the Seaton Delaval Brass Band made a presentation to the couple in appreciation of their service to the band. The members of Court Squire Bates Lodge made a presentation, in 1882, in appreciation of their kindness over many years. The Astley Social Club presented a dress ring to Mrs Dawson in 1885. In 1891 the Astley Touring Club gave three cheers to their president and his wife and in 1897 the Hartley Hastings Cricket Club presented a cigar case to Mr Dawson. However, it was at the death of Mrs Dawson in October 1895 that the communities regard for the Dawsons was most fully expressed. The cortege was followed by 36 mourning coaches and 2,000 mourners on foot, a visible symbol of the regard in which she was held by the whole community. It was in the Astley Arms that the various segments of the community interacted on a social basis.[87]

Conclusion

What insights can be gained into leisure practices in the villages? It is clear that the majority of the activities were associated with particular facilities, those influenced by the colliery officials, teachers and ministers. Additionally the majority exhibited a degree of organization. In fact, these activities provided a particularly warped view of life in the villages. At the same time, they all witnessed change in the 1880s – a critical era during which significant changes took place; in particular, entertainment and amusement became acceptable objects of leisure. It was also during the 1880s that there developed some community events at which the whole community got together, albeit in clearly defined roles – the flower shows and Children's Day. However, there was another dimension of leisure, which for the most part remains hidden from view. Yet this is a critically important element of life because it was so pervasive. These activities, of which we only get glimpses, were, for the most part, organized by the miners themselves. Music was central to life. This is attested to by contemporary comments on the pervasiveness of music in life from the brass bands, to the choral societies, to the sing-songs in the "pubs" to singing in the mine. Music was a fundamental ingredient of life. Music spanned the spectrum from symphonic music to popular ballads.

Another dimension of life that provided a focus for leisure activities was concern over conditions affecting their lives in the villages. In times of coalfield-wide strife miners gathered together to discuss their concerns. Perhaps more important to the individual villages was the evidence of their concerns over village life. In 1874, the inhabitants of Seghill must have spent a great amount of time and energy in organizing the campaign for a School Board. Similarly, in 1892, residents carried on a campaign over drinking water. These are simply the tip of the iceberg. These must have been an ongoing part of their lives. Finally, and perhaps the most hidden of all, were the ongoing efforts of the miners to improve themselves outside of managerial controlled initiatives. We got a brief glimpse of the Ashington Debating Society which for 30 years met weekly to discuss issues which, even today, would not be out of place in university seminars. The strength of this autodidactic tradition is difficult to gauge but one suspects it was quite strong.

Perhaps the more important question is to what degree were the leisure practices of the mining villages of East Northumberland similar to other working class groups? With respect to other coal mining communities across England, the activities were similar enough to support the idea of a mining culture. Nearly all colliery districts boasted brass bands, temperance societies, theatrical entertainment, flower shows, annual trips, and, most importantly, educational initiatives. The development of educational opportunities differed from village to village but were encompassed under one or more of the following: libraries, reading rooms, Mechanics' Institutes, literary and debating societies, the University Extension Scheme, the Workers' Education Association and a bewildering variety of lectures.[88] What held them together was the underlying drive for self-improvement. At the same time, most of these endeavours were characterized by a basic reality of mining leisure activities – impermanence – forms sprouted, flowered and then withered away. In fact, the differences were due to certain basic elements that influenced life in the mining districts: the role of the colliery owners and officials in the lives of the inhabitants, the physical proximity of the miners' houses to those of other groups, the composition of the population in terms of both where they were born and occupation, and in particular the role of the churches and chapels in village life. Yet, despite these significant differences it is possible to recognize a basic form of leisure activities that was common to mining communities across England.

Deeper insights can be gained by examining the provision for leisure activities within other working-class groups. In fact, many of the most popular activities crossed occupational and geographical lines. All the activities mentioned above found root in some working-class communities – "Rational Recreation" was common across the spectrum: brass bands developed, especially in the textile districts of Lancashire and Yorkshire, annual trips were institutionalized throughout England, temperance societies were formed in many districts, flower shows were popular amongst the working class and friendly societies proliferated throughout working-class districts of England. The intellectual world of the working class is brilliantly evoked in Jonathon Rose's *The Intellectual Life of the British Working Class* (2001). He demonstrates, clearly, the complex and pervasive nature of efforts at self-improvement. A few examples will suffice.[89] The immense popularity of Everyman's Library after its formation in 1906. This was built on many precursors, up to 100 between 1830 and 1906. The immense popularity of these series can be gauged from the fact that "By October, 1897 there were sixty volumes of Penny Poets with 5,276,000 copies in print, and about 9 million copies of ninety Penny Novels."[90] Reading tastes spanned the spectrum from the classics to "penny dreadfuls" and from literary magazines to the *Sporting Times*. Similarly miners' tastes in music were wide, from classical to brass bands to local songs. What Rose demonstrates is that there was an immense submerged autodidactic tradition amongst the working class wherever they lived.

There is one thing that emerges from an examination of other working-class districts: the importance of ownership of facilities. And it was in the larger towns and cities that conditions allowed for the creation of working-class-owned and operated facilities. From 1837, some of the larger friendly societies began constructing their own buildings. These were followed by Halls of Science (1839), Lyceums and "People's" Hall in Birmingham in 1846.[91] The creation of independently owned and operated facilities spread to London with the construction of working men's clubs. From the 1870s these clubs provided increasingly diverse activities which by the 1890s included pure entertainment.[92] This movement spread throughout England with the development of social clubs. Additionally, in Lancashire and Yorkshire in particular, during the second half of the century there was a significant increase in the number of working-class sponsored reading rooms. This led during the twentieth century to "one of the greatest networks of cultural institutions created by working people anywhere in the world" – the Welsh Miners' Libraries associated with the Miners' Institutes.[93] All these were significantly different from the facilities created by the middle classes: public parks, public baths, libraries and museums, although the degree to which they were utilized by the working classes is a moot point. More important to working-class life was the commercial provision for working-class leisure – pleasure gardens in London in the first half of the century and Gaff theatres from the 1830s.[94] While these were limited to London, music halls spread into the provinces. What this all suggests is that the working classes in different locations had a greater range of activities to choose from than did the villagers of East Northumberland. At the same time, most activities were available but in a different form, for example itinerant entertainers travelling through East Northumberland in the 1880s. However, by the second decade of the twentieth century most working-class areas had access to the epitomization of commerial entertainment – the cinema.

Finally and as a bridge to the next chapters we noted that one institution has been noticeable by its absence – the inns/public houses/taverns. While they briefly entered the picture in terms of annual celebrations, organization of flower shows, and hosting certain activities, they have, for the most part, been absent from the discussion. However, there is no doubt that, throughout the century, they were the most important social institution in the village – they dominated social life. The inns/pubs played a central role in the social lives of the working class across the length and breadth of Britain. In fact, they were even more pervasive in the larger towns and cities with a greater number of pubs/inns per head of population than in the mining villages. As Gareth Stedman Jones stated of the new working class culture that emerged during the period 1870 to 1900, "Its dominant institutions were not the school, the evening class, the library, the friendly society, the church or the chapel but the *pub*, the sporting paper, the race course and the music hall."[95] They became the centre of the social life of the community providing entertainment, meeting rooms, sponsorship and, in some cases, sporting facilities. As far as can be ascertained they were exclusively male domains. Females were not privy to the vibrant social life of the pubs.

3 The strength of tradition

Introduction

The leisure activities described in chapter 2 provide a particular view of the mining villages emphasizing the pervasive influence of the colliery officials, ministers and teachers. In fact, this view is a particularly biased and incomplete view of life in the villages. There was another dimension of village life that focused on sport. Most historians regard sporting activities as being trivial diversions from the real meaning of life and, in the final analysis, having no real significance.[1] They underestimate the role of sport in life. For many it provided the central focus to life and provided them with acceptable ways of expressing their independence, community solidarity, and their own masculinity. The history of sport reveals dimensions of life that are obscured by the emphasis upon morally uplifting activities sanctioned by the chapels, churches and colliery officials. This alternative culture, universally condemned by outside observers as a culture of drunkenness, brutality, and the pursuit of pleasure, the antithesis of middle-class culture, focused on the inns and innkeepers, the individuals and institutions universally reviled by the middle class. Yet the inns were the central social institutions in the majority of mining villages. Thus, we must turn to sport.

Sport cannot be looked at as a homogeneous whole. This chapter will deal with a group of traditional sports whose histories can be traced back into the eighteenth century and which remained visible until 1914.[2] While all were changed in subtle ways during the course of the century they were still played in 1900 in a manner that would have been recognizable to their forefathers 100 years earlier. Individually and collectively, they reflect the powerful forces of tradition and the conditions under which tradition thrived. However each was subtly different. Cockfighting, as we shall see, moved from being, arguably, the most popular sport at the beginning of the nineteenth century, to a sport that went underground and, thus, disappeared from view. Fox hunting, throughout, was the exclusive preserve of particular groups within East Northumberland. It was in rabbit coursing and shooting that the lines between the different social groups were crossed and became blurred. Quoits, potshare bowling and handball were played exclusively by miners. Pitch and toss was different from any others because it remained illegal throughout, thus, our view of the game is obscured by its invisibility except when it surfaced in the courts. Finally, unlike the non-sporting leisure activities the inns/public houses played a pivotal role in the sports.

Cockfighting

The history of cockfighting provides a rare glimpse into popular culture at the end of the eighteenth century. It also serves as a salutary warning against an oversimplification of the conditions of working-class life and the break between traditional and modern sports. The existence and popularity of cockfighting revealed a vibrant set of cultural practices, different from what was to come, while, at the same time, demonstrating some of the characteristics of modern sport: regularly scheduled competitions, an embryonic system of organization and generally accepted rules. At the turn of the century, cockfighting was the most popular spectator sport in Newcastle.[3] With a population of 34,000, Newcastle boasted seven cockpits which, from January to June, conducted daily "mains" except on Fridays and Sundays; far more frequent, in terms of the number of days in which it was offered, than professional football of the modern day. The pitmen of the north bank of the River Tyne played a pivotal role in all aspects of the sport. They were the primary breeders and trainers of the fighting cocks and were ardent spectators at the daily events. So important were they that on Pay Saturdays special matches were provided by the Newcastle pits. Cockfighting retained its popularity well into the century. In fact, during the 1840s, a time that according to historians had seen the abolition of the sport, a guardian of public morals in Newcastle launched a vehement attack against its existence in the city. Additionally, the Parliamentary Commission on the Employment of Children in the Mines, in 1842 commented on the popularity of cockfighting in the mining villages. As late as 1850 seven prosecutions indicated that it was alive and well. It was the magistrates of Newcastle, who in the 1850s closed the pits in Newcastle, that forced the sport underground. The decline and disappearance of the sport from the public domain was a result of external pressures and certainly not from within the mining communities. As the force of "Rational Recreation" and negative attitudes towards brutality to animals became more powerful the sport became the preserve of a small subculture. However, it never disappeared totally, surfacing on the odd occasion in Seghill in the 1870s, Waterloo in 1881, the Town Moor in 1893 and Morpeth at the turn of the century.[4] As late as 1892, the *Blyth Weekly News and Transcript* referred to "The first fight of the present season."[5] However, as a central focus of the miners' sporting life it was, to all intents and purposes, removed.

Fox hunting, coursing and shooting

Fox hunting epitomized pre-industrial, pre-mining England. Deeply rooted in the social structures of an agriculturally based society it maintained a robust presence until 1914.[6] Along with shooting, coursing and otter hunting, it provided the focal point of the social life of the landed aristocracy during the winter season.[7] At least five days a week were spent hunting, shooting or coursing. In fact, the social life at

Blagdon Hall was similar to that practised on the estates of the landed aristocracy throughout Britain. The first identifiable pack of hounds was formed, in 1818, by Sir Matthew White Ridley. This pack, located at Blagdon Hall, rode across the fields of East Northumberland until it was disbanded in 1846. Some idea of its pervasiveness can be gauged from an examination of the 1831–32 season which started on September 2, 1831. During the ensuing six months, the pack rode to the hunt on 77 occasions, at least three times a week. The pack ranged far and wide, mainly to the north and west of Blagdon Hall. However, they did venture into the coalfield on 11 occasions. During the season they killed 32 foxes, grounded 18 and lost 51 – in this year the foxes "won"![8] The Morpeth Fox Hounds were formed by John Cookson and located at his estate, Meldon Park. This pack can be traced with some accuracy from the 1870s until 1914. Fox hunting continued until the war, although the hunts tended to focus on the areas to the west of the coalfield. The hunts met increasing resistance as farming developed and the population increased. However, by 1912, the Morpeth Hunt rode on Saturdays, Tuesdays and Thursdays throughout the season from November until April 15, a total of 63 days. What had changed was the location of the hunts; they only brushed the mining areas at Longhirst, Bothal, Hartford House and Blagdon. Obviously the miners never participated except as spectators as the riders rode through the area. However, on one occasion, in 1912 during the strike, 13 miners turned up at the Braes of Derwent Hunt resplendent in cloth caps and mounted on pit ponies!

There were two other sports that were part of the winter season at Blagdon Hall: coursing and shooting. However, these differed from fox hunting in that they developed strong roots within the mining villages. And it was in these sports that the miners found allies in their battles against animal rights activists who attempted to have coursing and shooting banned.

Rabbit or hare coursing transcended class boundaries and brought individuals from widely differing backgrounds together in the pursuit of a common interest. Sir Matthew White Ridley, one of the most prominent landowners, was a strong supporter of coursing. So too was the Vicar of Seaton Delaval. They were joined by Joshua Rutter, the lessee of the Bebside Inn, owner of the Bebside Recreation Ground, and an active greyhound racer until his death in 1902. Miners, too, were avid supporters of the sport. However, their mutual interaction was limited to the prestigious hare coursing meets that were regarded as preparatory for the Waterloo Cup. In fact, coursing can be divided, roughly along class lines, into two different types, the more limited hare coursing that became part of the social activities of the upper echelons of society and the more popular rabbit coursing whose support was drawn from the mining villages.

Hare coursing with greyhounds was, in all probability, practised throughout the century. Morpeth Coursing Club, patronized by the Earl of Carlisle, was active between 1841 and 1851. It then disappeared from view and was replaced, in the mid-1850s, by the Bothal and Gosforth meets. These meets, held in the spring and autumn, focused on the Duke of Portland's Bothal and Gosforth Park estates. The nature of these meets was exemplified in the October meet at Bothal in 1871.

The meet consisted of two events, the Bothal Derby which attracted 157 entries at £2 per entry and the more prestigious Longhirst Oaks with 177 entries at £3 10s per entry. The prize money was over £100 for first place in each event.[9] The social composition of the participants crossed class lines and brought different social groups onto the same field of competition. As early as 1871, the Bothal Club specifically organized a meet for the "Tenant farmers and pitmen on the estate."[10] This was followed, in 1872, with "An all age stake for the miners of Ashington Colliery."[11] The 1880s witnessed the expansion of the number of meets held in spring and autumn. It was during the 1890s that the meets expanded to include an event for greyhounds that were the "Bona-fide property of Working Men."[12] Thus, a limited number of working men did participate in the events. However, their social position was explicitly recognized with votes of thanks on behalf of the working men "for allowing them to participate."[13] These greyhound meets were, for the most part, patronized by landowners, farmers and several publicans. Publicans frequently served as the secretaries of the meets. However, coursing did serve as a link, no matter how tenuous and fragile, between the different levels of society. The relationships were rooted in old England and not the new.

What was rabbit coursing, how did it develop, and what were its characteristics? In November 1891, Lieutenant Colonel W.L.B. Coulson reported to the weekly gospel temperance meeting in Central Hall, Newcastle on what he termed "A vicious exhibition."[14] On walking through Ryton Willows he happened on a group of pitmen from Blaydon engaged in the popular sport of rabbit coursing. He provided a vivid account of a group of miners with their whippets held on leashes. The rabbits were brought in a sack. Two miners stepped to the starting line with their dogs held on a leash. A rabbit was let go and allowed to run for approximately 25 yards at which time the whippets were unleashed. The dog that killed the rabbit was the winner. This was rabbit coursing, a sport with deep roots in the mining villages.

Prior to the 1860s there is only fragmentary evidence on the extent of rabbit coursing. However, even from this evidence, it is evident that miners tested their dogs against other dogs both in coursing and straight racing. From the 1820s these matches and courses were held on public land, in particular, on the Newcastle Town Moor, Newbiggin Moor and friendly farmers' fields. It is not until the 1860s that it is possible to discern certain basic characteristics: the basic forms of competition, challenge matches between individual miners and their dogs, sweeps or handicaps which attracted a number of dogs to a particular competition, and the role of the inns and innkeepers. Coursing was held on an infrequent basis in Fenham Running Grounds, on the outskirts of Newcastle, after it opened in 1862. Perhaps more important to the development of rabbit coursing, within the villages, were the periodic matches, handicaps and sweeps, all sponsored by local innkeepers. This was the first evidence of the process of commercialization that was to engulf sport in the coalfield. As in many other sports the early 1870s, the high point of prosperity in the century, witnessed a significant increase in interest in rabbit coursing. This lead to the opening, in January 1876, of the first commercial rabbit coursing ground, the short-lived Killingworth New Grounds.

Plate 3.1 Rabbit coursers, Ashington.
Reprinted with permission from Harrison, W. (1990) *Ashington: A History in Photographs*, Morpeth: Northumberland County Library

In fact, 1876 is an ideal year to examine the nature and development of rabbit coursing and dog racing. During the year, 85 events were held involving greyhounds (6) or small dogs (79) in straight racing and rabbit coursing matches and handicaps.[15] These events took place at 17 different locations, grounds adjacent to inns, during the coursing season from September until April. They competed for a not unsubstantial sum of £1,224. However, inns were involved in different ways. In the case of the 33 matches they were used in the traditional way as centres for meetings leading up to the contest. More indicative of the future were the 52 handicaps. All handicaps were sponsored by innkeepers and represent the commercialization of the sport. In fact, 38 of the contests were held at one location, the short-lived Killingworth Grounds. The grounds were opened on January 1, 1876 by two innkeepers, and were devoted entirely to dog racing and rabbit coursing.[16] This reflected the explosion of interest in dog racing in its various forms and it evoked an immediate response from the authorities. On June 12, 1876, the two innkeepers were fined £20 each for an offence under the betting act.[17] By the end of August all commercial grounds in East Northumberland had been closed. The sport received another blow in 1880 when Newcastle Town Council closed the Town Moor to rabbit coursing, potshare bowling and shooting.[18] This, of course, was not the end of rabbit coursing, it simply moved within the coalfield. The subsequent history of rabbit coursing and dog racing demonstrated increased commercialization. Even though individual innkeepers still sponsored events the majority were held in four commercial grounds that opened around the turn of the century. Between September 1912 and April 1913, 91 per cent of all competitions were held in these grounds. In the same season, 108 of the 121 competitions were handicaps.[19] The two dog matches with their limited appeal for ground owners gave way to handicaps and sweeps whereby a number of dogs, usually 16, 32, 64 or 128, engaged in a number of knockout rounds before a winner was finally declared. Thus, for a small outlay entrepreneurs attracted a relatively large number of dogs, guaranteeing increased competition – an event that attracted 128 dogs provided 126 matches – a good afternoon's entertainment and, more importantly, increased opportunities for gambling. In fact, the commercialization had gone a step further with each ground staking out its own particular focus and, thus clientele. Allotment Ground was the focal point of greyhound racing while the nearby Robin Hood Inn and Shankhouse promoted small dog racing. The differences were clear, particular grounds appealed to particular segments of the dog loving fraternity. Specialization had come to characterize rabbit coursing. In fact, the history of rabbit coursing reflects basic processes that characterized sport in general: the creation of a recognized season in the 1870s, commercialization and centralization by the turn of the century, and specialization by 1913. In all probability whippet racing and rabbit coursing appealed to a small group of devotees. This is supported by the fact that only 3 of the 17 miners interviewed during the 1970s mentioned these sports.[20]

The history of rabbit coursing raises a fundamental question: why did the sport, which was severely condemned by newspaper editors, ministers of the cloth and animal rights activists, survive and, indeed, prosper? It first serves to illustrate the limitations of the power of the authorities, ministers, and the self-appointed guardians of public morals. Even though vast amounts of newspaper print were devoted to promoting the abolition of rabbit coursing and other "brutal" sports, the final result was that they failed to change life within the villages. However, there is a reason why all efforts failed – interest in coursing crossed class lines. In 1895, the Newcastle Branch of the Sporting League was formed to fight attempts to abolish stag hunting, shooting and rabbit coursing.[21] Its membership read like a Who's Who of Northern gentry – the Earl of Ravensworth, Earl Percy, Lord Warkworth, and many others. Within the coalfield, the sports were actively supported by such luminaries as Sir M.W. Ridley. The representatives of the aristocratic landowning class were strong supporters and participants in these country sports and thus acted as a bulwark against calls to eliminate the sports.

Shooting was another sport that transcended class boundaries. However, shooting never crossed class lines. It, along with fox hunting and coursing, provided the focal point for the winter social season. At Blagdon Hall shooting filled the days from September until the end of January. Between September 1, 1869 and February 11, 1870 over 10 days a month were spent shooting on the Ridley lands surrounding Blagdon.[22] On the cool mornings of November and December from one to five guns gathered at one of 23 locations, all within a mile or two of Blagdon Hall. During the 1871–72 season 80 pheasants, 724 partridges, 62 snipes, 13 woodcocks, 39 wild fowl, 201 hares and 206 rabbits were shot for a total bag of 1,410. The game shot on the 61 days was distributed to farmers, estate workers and patients in the infirmary. The distribution of game served to reinforce the social relationships that flowed from the Ridley Estate: hares and rabbits to the workers and the partridges and pheasants to farmers and local gentlemen. Little changed during the next 40 years. Shooting still dominated the period from September to January. Although the number of shooting days had declined to 23, the number of guns had increased and the total bag for the year was 1,729 of the same game, although the most plentiful were the 973 pheasants.[23] The individuals invited to participate in the shoots were drawn from the highest echelons of Northumbrian society: the Burdons of Hartford Hall, the Joiceys of Longhirst Hall, the Cooksons of Melton Park and F.J. Snowball of Seaton Burn House. There was only one other shoot that was the social equivalent of the shoots at Blagdon Hall. It took place on the Estates of the Duke of Portland on the north bank of the River Wansbeck between Morpeth and Ashington. However, it was not the Duke of Portland, although he maintained the right to shoot at any time, but the resident agents, the Samples, who shot. During 1913, W.C. Sample shot on 21 occasions.[24] Shooting game, however, reached deeper into the society than the very upper levels. For example, the rights to shoot on the 19,000 acres of the Bothal Estates were rented out to 18 other individuals. For a small but influential segment of Northumbrian society, shooting was an important wintertime activity.

Shooting was not the exclusive preserve of the upper classes of East Northumberland. It has a long history amongst the miners. It was the enactment of the Game Laws in the 1820s and 1830s and the consequent restriction of shooting rights that forced the miners to move onto specific locations and thus, once again, the innkeepers moved to centre stage. Throughout the century certain things did not change – the role of the inns and innkeepers, the forms of competition; infrequent matches and the more popular sweeps and handicaps; the birds to be shot, sparrows and pigeons, and the fact that there was a significant difference between sparrow shooting and pigeon shooting, Pigeon shooting was more prestigious, attracting the larger prizes in the matches, on occasion reaching £100.

Although irregular matches and handicaps did take place during the 1840s and 1850s it was in the 1860s that a more regular programme of competitions emerged. Even then it is difficult to identify any logical patterns. What emerges is a picture of an ever-shifting focus on different inns with up to seven inns promoting competitions in a given year. One of the most interesting differences from the majority of miners' sports was the difference in the prizes awarded to shooters in matches, on the one hand, and sweeps and handicaps, on the other. While the challenge matches were always for money stakes, those offered for the sweeps and handicaps were always in kind rather than money – for example, firkins of ale, wheelbarrows, silver watches, eight day clocks, various types of guns and cases of stuffed birds. It is apparent that these prizes related to traditions in the mining society that have been obscured. However, the challenge matches between individuals were always for money prizes. A major shift in terms of prizes occurred in the 1870s; prizes such as those mentioned above disappeared and all shooting contests were for money stakes or prizes. As with nearly all sports in the mining villages, the 1870s were a watershed between the old and the new.

Until 1888, a limited number of inns (26) sponsored matches and handicaps on an irregular basis. From 1888 to 1900 there was an explosion of interest in shooting; 82 inns sponsored shooting competitions.[25] Prior to the 1890s, shooting lacked any formal organization, depending heavily upon the whims of individual innkeepers. For a brief time during the 1890s, the sport acquired some of the characteristics of modern organized sport: the development of regularly scheduled competitions, a degree of organization, and the formation of formally constituted clubs. During this period 10 inns promoted the sport on an annual basis and came to be regarded as the shooting centres of the district. In 1898, they sponsored 66 of the 99 handicaps promoted in the district. All had certain characteristics in common: the innkeepers were enthusiastic participants in shooting; handicaps were sponsored on a regular basis; all boasted Gun Clubs; and all had open space available adjacent to the inn. During the 1890s, competitions were available on a regular basis throughout the season. The inns that sponsored competitions on an ongoing basis were those that organized gun clubs in conjunction with the inn. Between 1887 and 1913 38 gun clubs were organized. The majority failed to last a year. There was a small group of four that attained a degree of stability, and it was these clubs that held regular gun club shoots from September to April. In some cases they had fortnightly competitions and frequently hosted prestigious challenge matches. The membership of the clubs rarely exceeded 30 members. It is estimated that there were approximately 1,200 shooters in East Northumberland. By 1914 all the clubs had disappeared and shooting contests had become a rarity.

It had become a minor sport with a core of ardent supporters. Again, this is supported by the fact that none of the 17 miners mentioned shooting.[26]

The actual contests provide a partial view of the sport and of sports in general. All sports required a degree of organization in order to ensure successful competitions. Shooting provides a glimpse into the infrastructure that developed around many of the sports. The nature of this infrastructure is illustrated in a pigeon shooting handicap which was held on January 27, 1896.[27] The handicap, for a prize of £25, was sponsored by three innkeepers. Their inns were the venues for a series of meetings at which competitors deposited their entry fees and handicaps were arranged. The three innkeepers made arrangements with Blyth Council to secure the use of a portion of Blyth Links for the shoot. George Thornton, of Bebside, a professional supplier of birds, was hired to provide the birds and to frame the handicap. J.R. Thornton was hired to act as trapper, the person who released the birds. The most important task of keeping the books was placed in the hands of William Nicholson, also of Bebside. This meet, which attracted several hundred spectators, was typical. It illustrated all the necessary requirements: prize money, location, a supply of birds, a core of administrative personnel, and opportunities for gambling. The innkeepers played the central coordinating role, provided the prize money, secured a location, and hired the key personnel. For the most part shoots were held in fields adjacent to the inns, in friendly farmers' fields, on public ground and, increasingly in the twentieth century, in commercial grounds.

From the earliest years, the highest level of shooting was to be found in the challenge matches that pitted two individuals against each other. Any history that ignores this highest level does an injustice to the history of shooting. From the 1860s it is possible to trace the various contests and championships that were contested for. From the 1880s a year never passed without at least one challenge match. At the same time they were swamped, in terms of numbers, by the popular sweeps and handicaps. Between 1889 and 1912, there were, on average, 10 challenge matches a year. As with the handicaps all matches were associated with the inns. Perhaps a glimpse into the world of challenge matches can be gleaned from the history of George Hannay of Bedlington between 1900 and 1910.[28] During this period he was the premier shooter in the district and was acclaimed "World Champion" in at least two different events. In a three week period in 1901 he competed in two matches which brought him £100. He won the so-called Championship of the World and £100 at a contest in Newcastle. Following this victory he traveled to Dipton, County Durham to shoot for the One Barrel Championship of the World and £50 – he won. He did not limit himself to one form of shooting, participating in contests with different guns and with different objects – pigeons, sparrows, clay pigeons and glass balls. Although it is difficult to prove, it is likely that Hannay was, indeed, a professional shooter.

To what degree was the experience of East Northumberland replicated throughout Britain? The social practices of the inhabitants of the country houses of East Northumberland were replicated in country houses throughout Britain. The winter social season was created around fox hunting, shooting and coursing. What the history does suggest is that the practices of the elite were more resistant to change than those of other segments of society. However, more interesting because they have received little attention, was the popularity of dog racing, coursing and shooting. With respect to coursing, when one moves from greyhound coursing to small dog rabbit coursing the picture is unclear since little or no evidence as to its form and popularity exists. However, the popularity of dogs in working-class communities throughout Britain is suggestive. Pigeon shooting, on the other hand, was participated in by all segments of society from the Aristocratic Pigeon Shooting Club (1868) to the tavern based contests in the Black Country.[29] In Central Scotland during the 1870s it was "One of the most fashionable pastimes."[30] However, there were differences based on class. Both in Scotland and England clubs developed in the 1860s and 1870s. These were, for the most part, upper and middle class. And it was these clubs that bore the brunt of the campaign to abolish pigeon shooting that culminated, in 1883, in a discussion in Parliament. This was the beginning of the gradual decline of pigeon shooting. The upper and middle classes turned to glass ball and clay pigeon shooting. In East Northumberland and the Black Country there is little evidence that the tavern based pigeon and sparrow shooting was affected at all. In fact, the highpoint of shooting in East Northumberland was the 1890s. Of course, sparrow shooting was a working-class sport. What the limited evidence suggests is that the values of the upper and middle classes did not reach very far into working-class life and activities. Whether this illustrates the power of tradition or the lack of interest of the upper and middle classes in the realities of working-class life is a question that has yet to be answered.

Potshare bowling, handball and quoits

In considering the histories of potshare bowling, quoits and handball we move deep into an area that, for the most part, has remained hidden from view, the lives of the unrecognized and unheard from majority of miners, those who eschewed education, the chapel, and morally uplifting activities. It was games such as these that were deeply rooted in the community and stretched back well into the eighteenth century. And it was these games that defined the community and which have, to all intents and purposes, disappeared during the twentieth century. They provide insights into the nature of the mining communities and the importance of sport in the lives of the miners. That potshare bowling and quoits could, at the outbreak of the First World War, claim to be the most widely played and watched sports attests to the power of tradition in the face of the massive changes of the nineteenth century.

The history of handball, also known as fives, illustrates the barriers facing anyone attempting to obtain a glimpse into the hidden world of mining life, the world of the majority. That handball was a widely played sport is attested to by the fact that just prior to the war, when references to the game had disappeared from the newspapers, it was still being played throughout the coalfield. In Ashington, one street would challenge another to a handball match.[31] For the more important games the gable end at the recreation ground was the recognized location for matches that attracted players such as Jack Mordue, the outside right for

Sunderland from 1908 to 1920.[32] About seven miles south at the Hastings Ball Alley in Seaton Delaval, handball was, according to Jimmy Tabarar, played in afternoons, evenings and weekends. There were always battles on Saturday afternoon when the protagonists stripped to the waist and played with bare hands. Because there was betting there was always a referee.[33] Jimmy Tabarar's claims were substantiated by Huey Smith, who talked about handball at the Hastings Arms and the use of wire netting to stop the ball rolling into adjacent gardens.[34] Handball was alive and well at the outbreak of the war, by which time competitive handball had disappeared from public view. In fact, it continued to be played in the adjacent Durham coalfield until the 1950s.[35] It is important to emphasize that the high level competitive challenge matches and handicaps reveal only the tip of the iceberg and were built upon a base of participation within the villages.

Although there were 22 ball alleys spread across the coalfield there was one particular area of concentration and it is from this area during the period 1883 to 1901 that some insights into the form and function of handball can be gleaned. There were six ball alleys, within a two mile radius of the Astley Arms, that hosted relatively regular competitions. Handball entered a new era in 1886 when James Dawson of the Astley Arms built a ball alley at his inn with seating for 600 spectators.[36] A similar initiative was undertaken in nearby New Hartley where the newly opened Hartley Arms also provided a facility for spectators and players. This ushered in a flurry of activity. Between September 25, 1886 and December 18, 1886, matches and handicaps were held every Saturday. These contests, which pitted individuals or pairs of players against other individuals and pairs, attracted large numbers of spectators even when the weather was "inclement". Single- or double-handed matches were to either 21 or 33 points. The articles of agreement, signed before the match, laid down the conditions of the competition and identified the referee and stakeholder. It was the nature of the competition at Seaton Delaval that revealed a probable cause for the lack of interest in handball by the media. During the period 1886 to 1901 the majority of competitors were from Seaton Delaval, New Hartley or Cramlington. Only on two occasions did players from outside the immediate area compete.[37] Thus, it appears that handball was a local concern with pockets of interest across the coalfield.[38] Unlike the neighboring Durham coalfield where handball was a major sport that, on occasion, attracted as many as 3,000 spectators, handball in East Northumberland never developed the intra-coalfield competition.[39] This was evident in the lack of "Championship" matches, which were the focus of attention in quoits and potshare bowling. Like all the traditional sports, handball epitomized what it was to be a "man" – strength, endurance and perseverance. Handball remained rooted in the communities where it was played until the war. However, the history of the game at a competitive level reveals a lack of regularity.

While the history of handball remains, for the most part, obscured, the same cannot be said about what was, arguably, the most popular and high profile sport throughout the nineteenth century, potshare bowling. If there was a single sport that demonstrated both the immense power of tradition and, paradoxically, the inexorable nature of change, it was the game of potshare bowling, a game

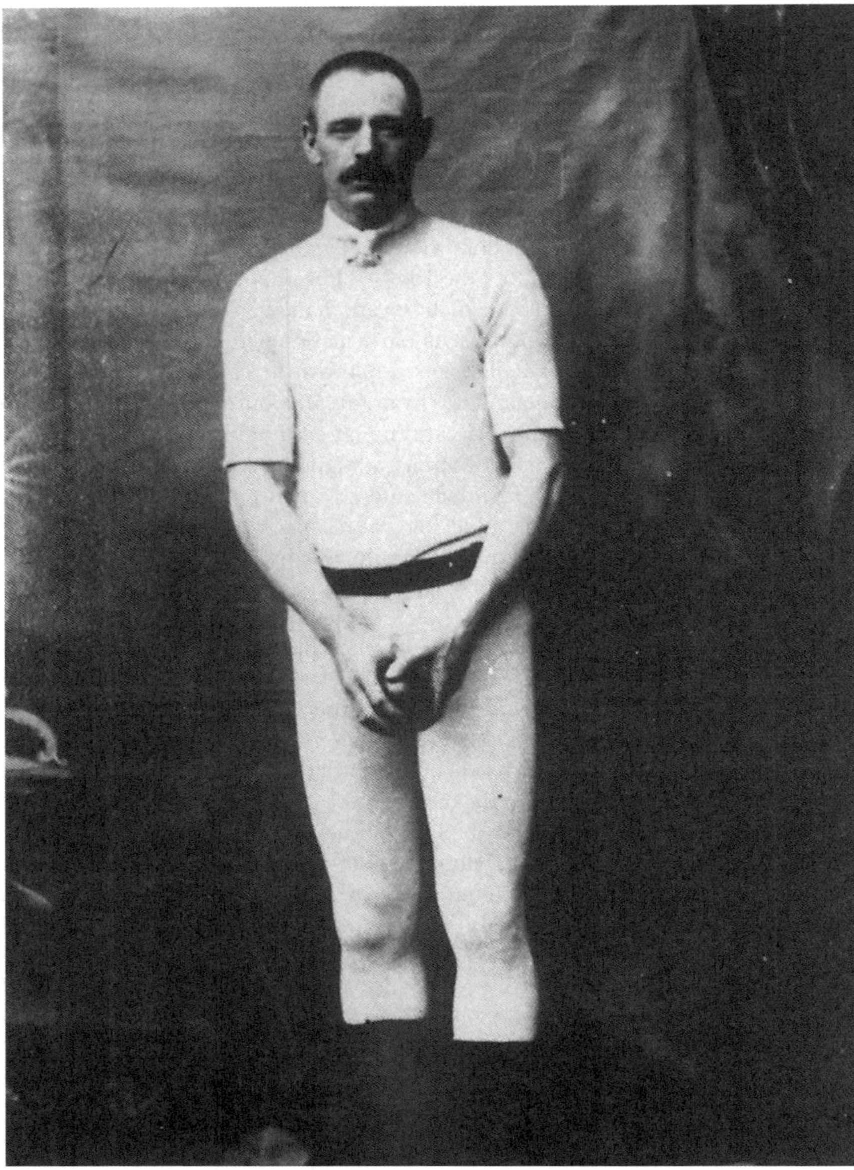

Plate 3.2 Potshare bowler.
From the private collection of Mike Kirkup

played solely by the miners of Durham and Northumberland. It is possible and, indeed, probable that in 1914, potshare bowling and quoits attracted more players and spectators amongst the miners than did the game of football.[40]

What then was the nature of this game that so dominated the miners' sporting lives? Basically it was a simple game that pitted two men with potshare bowls, weighing between 5 and 50 ounces, against each other over a predetermined course, usually called the "mile". After alternating throws, the miner who got to the end of the course first was the winner. In many ways the game did not change during the course of the century – the great grandfathers of men playing in 1900 would have felt comfortable with the game as it was played. In the first place, the game was played on public land; 91 per cent of all contests were played on Newcastle Town Moor and Newbiggin Moor. From the 1820s, the Town Moor was the recognized centre of championship bowling. Unlike modern sports with their pre-determined league schedules, there was no regular schedule of events, the occurrence of a match being dependent upon the number of challenges or the sponsorship of handicaps. Thus, the frequency of matches varied widely. Additionally, the basic competitive format remained the same, being either challenge matches or handicaps. There were no written rules or formal organizations. Matches were arranged by following generally recognized procedures. The initial challenge, if local, would be done by word of mouth or, for the more important matches, after 1858, by placing the challenge in the *Newcastle Daily Chronicle*. The two combatants and their supporters would meet at a designated inn to formalize the match. At that meeting and subsequent meetings they would discuss handicaps, location of the contest, the referee, stakeholder, stakes and the date of the match. In fact, the actual match was the end point of a complex set of negotiations involving relatively large groups of people. In the important championship matches the two protagonists would go into training for periods of up to six weeks. There was no season with matches being played throughout the year. Additionally, age was not a barrier to competition with several champions competing for over 30 years. Since the matches were played on public ground entrance money was never charged. This meant that crowds from a few hundred to over 20,000 that flocked to the matches did not have to pay to watch their champions perform. At the same time, *all* matches were played for money stakes, frequently as high as £200. Paradoxically, and difficult to understand, there was little concern over the conditions under which the game was played – a level playing field was not a priority. Rarely were games cancelled because of inclement weather. "Awful conditions", "miserable weather" and "wretched conditions" discouraged neither spectators nor contestants. Snow, thunder and lighting, and torrential rain failed to dampen their ardor. On February 10, 1902 the local football derby between Newcastle United and Sunderland was postponed because of snow.[41] Less than 2 miles away, on the Town Moor, bowling continued as usual. This lack of concern over conditions was rooted in mining life and their acceptance of the role of "luck" in both life and games.

Potshare bowling also illustrates the complete apathy of public authorities to the miners and their lives except when they became visible, in other words, when the paths of the miners and the citizens of Newcastle crossed. This occurred several times during the century. In 1805, 1822, 1871, 1874 and 1880 Newcastle Council attempted to remove the bowlers from the Town Moor, all to no avail. In fact, the attempt made in 1880 illustrates that miners were not powerless in the face of authority. In January 1880 the magistrates issued a total ban on bowling on the moor. Within two weeks the moor was devoid of action. By 1881, the miners were back on the moor bowling where they had bowled for the past 100 years. This was a result of two factors, the first was that the miners fought the Council in court on the grounds that they had been denied access to public land. The Queens Court in London found in their favour. However, even before that the Council had succumbed to demands made, not by the miners, but by a deputation of businessmen representing 1,200 large rate payers who complained at the loss of business consequent upon the demise of bowling – the miners' wives were no longer shopping in town. The miners were not powerless against authority and, in fact, vigorously defended their own interests.

Potshare bowlers were the undisputed heroes of the coalfield. Perhaps their importance was reflected most clearly in the early 1900s when the champions of the day were frequently measured against the champions of the past: Davy Bell of Benton in the 1820s, Harry Brown of Gateshead, in the 1830s and 1840s, Tommy Saint of West Moor (1850s and 1860s), John Gibson, Pegswood and Robert Gledson, Dudley in the 1870s, and Robert Nicholson of Choppington in the 1880s.[42] These names were transmitted from generation to generation through conversations in the pits and pubs. The young boys learned from their fathers the traditions that were important to them and thus bowling remained part of a living culture.

Despite the fact that the game of 1900 was very similar to the game of 1800, significant changes had taken place, changes that spelt the demise of bowling in the inter-war period. The most important of these changes was that match bowling declined and handicap bowling increased. With respect to match bowling there were two distinct eras, pre-1869 and post-1869. Prior to 1869 bowling was ever-present, but not very popular, averaging between 3 and 30 matches a year. In the six years from 1869 to 1875 the number of matches increased from 20 to 240. This ushered in the great era of match bowling, which extended to the 1890s. After 1890 there was a gradual but perceptible decline in match bowling. However, even in the great era of bowling it was characterized by massive annual fluctuations in the number of matches. In fact, the ups and downs were symbiotically related to economic conditions within the coalfield and could be used as a barometer of economic conditions. Matches were replaced by handicaps, which reached their peak in the period 1903 to 1910. The emergence of handicaps as the dominant form of competition was reflected, most clearly, in the days devoted to each.[43] In 1876 there were 80 days when matches were played and only 12 days when handicaps were played. By 1905, this had changed totally. While 83 days witnessed matches on the moors and beaches, 176 days were devoted to handicaps. And it was the local handicaps that illustrated the popularity of the game. For example, a bowling handicap in Seaton Delaval in November 1898 attracted 41 participants from the immediate locality.[44] The whole focus of bowling had changed. The question is why?

The answer to the question lies in the changing role of the inns in the promotion of potshare bowling. Prior to the turn of the century, the majority of handicaps were sponsored by inns located within the coalfield. Between 1886 and 1913, 38 inns within the coalfield sponsored handicaps, but rarely more than once. However, it was in Newcastle that the real changes took place. In 1904, the innkeepers of Newcastle decided to reduce the entrance fees from 5 to 2 shillings. This was the beginning of a concerted effort to promote bowling on the Town Moor. Between 1904 and 1908, 24 inns sponsored handicaps. There was more activity on the moor than there had ever been. Additionally they made an effort to attract new competitors and spectators from outside the mining villages – novice handicaps, handicaps for glassmakers and other working-class groups. Events were promoted on weekday evenings and Wednesday afternoons. Prior to 1900 over 95 per cent of the bowlers were miners. By 1913, over 40 per cent came from villages with no coal mines. By 1912, the experiment of the Newcastle innkeepers had failed and the focus returned to within the coalfield. However, the attempt to commercialize the sport by the Newcastle innkeepers had destroyed the traditional foundations of the sport. It was the miners who provided the players, referees, organized matches and were involved at all levels of the game. It truly was a game operated, nearly totally, by the miners. They used the inns as locations to make matches, deposit stakes, negotiate conditions and socialize afterwards. It was the attempt by the innkeepers to exclude the miners from running their own game that helped lead to the demise of bowling.

While potshare bowling was, undoubtedly, the most prestigious sport amongst the miners, there was another sport that reached even more deeply into their culture – quoits. It was not only a competitive sport for stakes, but one that was enjoyed and played by large numbers. However, it is extremely difficult to get any sense of its development except through the prism of high level competitive quoits. Its history remains blurred and it is difficult to get a clear view of this sport that so dominated their lives. In many respects it was similar to bowling – the dominant forms of competition were challenge matches and, later in the century, handicaps. The inns played a similar role acting as the centre for making matches, undertaking the negotiations and having post-match celebrations. Most of the matches were for money stakes and gambling accompanied nearly all the matches. The basic development was the same – evidence of quoits reaching back into the eighteenth century, spasmodic evidence until the late 1850s, increasing visibility in the 1860s and the beginning of more extensive competition in the 1870s. The 1890s were the highpoint of quoits although it maintained a strong presence right through to the war. It also provided sporting heroes for the villages and the coalfield as a whole. However, its history is somewhat different from that of potshare bowling and it is these differences that offer different insights into the nature of life in the mining villages.

Quoits was, and still is, a relatively simple game that required little more than a stretch of grass or clay. The game was played with round quoits on a piece of ground, either grass or clay, with two hobs located between 18 and 25 yards apart. The rules and conditions of quoiting were never standardized. It was this simplicity that made the game so appealing; miners played wherever there was spare ground. From the earliest days the inns provided these rudimentary facilities and thus became not only the site for making matches, but the actual site of the matches themselves. What better game for the innkeepers to promote, than one where everything focused on the inn? Prior to 1876, the main challenge matches focused on the commercial grounds in Newcastle, run by innkeepers, of course. There were spasmodic handicaps and matches held at inns on an infrequent basis. The prizes for the challenge matches were always money stakes. However, the prizes for the majority of handicaps were such things as eight day clocks, silver watches, pairs of boots, hams, sacks of flour and legs of mutton. The closure of the commercial grounds in 1875 and 1876 forced quoiters into the coalfield. Due to deteriorating conditions in the coalfield leading to the strike and lockout of 1877–78, the sport did not flourish at first. This was soon to change. During the early 1880s there was an explosion of interest in quoiting amongst the innkeepers in the district. A small number of innkeepers built covered grounds in order to attract spectators. By 1884, there were 89 contests promoted by 33 inns.[45] More importantly there had been a complete shift in the nature of the prizes, nearly without exception the handicaps were now for money prizes, a total departure from the old traditions.

The 1880s were simply the prelude to the expansion of the 1890s. In particular the number of competitions increased dramatically. By 1893 there were 192 contests offered at 31 inns spread throughout the coalfield.[46] However, there were clearly identifiable centres of quoiting run by innkeepers who recognized the commercial potential of the game. It was these innkeepers who built facilities to attract quoiters and spectators. This witnessed the increased commercialization of the game with the sponsorship of regular matches and handicaps and the building of covered quoit grounds with seating for spectators. James Dawson of the Astley Arms, Seaton Delaval created his first quoit ground in 1882. However, it was in the mid-1890s that he built a shed to house spectators at the quoit matches and handicaps. In fact, 50 per cent of the competitions, in 1893, were held at five inns. While contests were held throughout the year, it was during the 1890s that a recognized season emerged, roughly May to October. By 1913, the top five inns now accounted for 76 per cent of the contests and the focus had shifted to the Blue Bell, Shiremoor, the home ground of the recognized champion Will Summers.[47] At the outbreak of the war, commercial quoiting was clearly focused in a relatively small number of inns. However, it should be recognized that the commercial, competitive quoiting was based on a foundation of mass participation. The change in terms of participation was based in the large changes that took place in the early 1890s.

While innkeepers were commercializing the game, other events precipitated quoiting into its greatest era, one in which it truly became a sport involving mass participation. Sometime in the late 1880s and early 1890s, the colliery owners and managers changed their views on what was acceptable in the Mechanics' Institutes. They accepted that the institutes could include both education and amusement. From 1890 nearly every Mechanics' Institute included a quoit ground. Immediately the miners started promoting handicaps. One of the earliest was held in Choppington Mechanics' Institute in 1893.[48] It attracted over 100 competitors and was held over June, July and August. This was simply the first of many, as the miners rushed to participate in their own games. Throughout the villages of East Northumberland quoits began to be played on a nightly basis in the friendly climes of their own institutes. This process escalated in the early 1900s with the development of miner owned and operated social clubs. Nearly without exception these clubs included a quoit ground that was used on a regular basis. Their regular handicaps attracted large numbers of members. For example, in 1908 Burradon Workmen's Institute's quoit handicap attracted 128 competitors.[49] It was concluded within one week. Later in the same year Dinnington Colliery Literary Institute's handicap attracted over 300 members.[50] As Jimmy Tabarar of New Hartley said of the pre-war period, "It was played every night of the week."[51] It was within the context of these clubs that quoits moved, hesitatingly, into the world of modern sport.

It was in the early 1900s that the first moves to provide inter-club competition developed. In 1906, Seaton Terrace Workmen's Club hosted Annitsford Workmen's Club in the first reported inter-club competition. This was followed in 1907 by two separate initiatives that reflected a distinct departure from the traditional ways of competing. Pegswood Social Club and Pegswood Institute, located on the north western edge of the coalfield, joined five Morpeth clubs to form the Morpeth and District Quoit League. This league of colliery and non-colliery clubs was the only league with any degree of permanence. In 1908, a Northern Clubs Quoit League was formed under the auspices of the Northern Workmen's Clubs.[52] This five team league survived until 1910. The weakness of its base is illustrated in the fact that teams from 17 different institutes played during the league's three year existence. Interestingly, these leagues differed from the matches and handicaps in that the prizes comprised cups, pipes and other things, not prize money, which was fundamental to the matches and handicaps.

What has been missing from this account, so far, has been the individual champions who became local heroes. However, unlike with potshare bowling, it is difficult to get any clear view of who these individuals were. One Will Summers of Backworth, who was claimed, by his contemporaries, to be the champion of England for 17 years, is visible only for brief moments. Some idea of the nature of their careers, the amount of money involved, and the nature of the whole process can be gleaned from the history of William North, a hewer at Dudley Colliery who competed in quoit challenge matches throughout the 1890s.[53] An idea of the life of a "professional" quoit player can be gauged from his involvement during the period May 31 to August 22, 1898. North, who challenged some of the leading players, was not, himself, at the top level of quoiting but this brief overview of three months illustrates the popularity of the game. During the period North played five matches involving prize or stake money totaling £370, more than three times the annual salary of a hewer. This schedule of matches involved a number of meetings at which conditions of the match were negotiated, the size of the stakes decided, the locations and dates for depositing the stakes arranged, the referees agreed upon, and the handicaps decided. North and his sponsors were heavily involved during June, July and August. On May 31 they met with R. Taylor of Cowpen at the Turks Head Inn, Newsham to negotiate a match. An agreement was reached but the match was declared off when a dispute arose over the depositing of stakes. This was common in match making, more never made it to the quoit pitch than did. On June 2, 1898 North lost a match for £10. This match was played at the Turks Head, Newsham, one of the quoiting centres of the district. Five days later North won a match for £20 at the Victoria Grounds, Newcastle. After the conclusion of this match, the competitors and their sponsors adjourned to the Victoria Arms Inn, Newcastle to organize a rematch for £20 a side. North won this match which was played on the following Monday. Thus in less than two weeks North had competed for a total of £70, a not inconsiderable sum of money. The referee at the last match was Harry Snaith, mine host of the Patent Hammer Inn, Gateshead and one of the leading quoit players in the north. North challenged Snaith. After several meetings, articles of agreement were signed and placed with Chris Barrass of Newcastle for a match at the Victoria Grounds for £50 a side. The final deposit was paid, just prior to the match, at Grahams Arms, Scotswood Road. After losing the match, North immediately challenged Snaith to a rematch for £200. The challenge was not accepted. What this brief history of William North's quoiting career in the summer of 1898 indicates is the complex nature of the interactions surrounding a quoit game. The contest itself was the termination of a complex set of social interactions involving a number of people in various places. During the three month period North and his supporters travelled to Newcastle on at least nine occasions to play or negotiate matches. This was nearly a full-time job.

Quoits pervaded the life of the community in ways that other sports did not. From the 1850s it was an integral part of communal celebrations. Quoit handicaps were a part of numerous annual sports and community events. Additionally, the game was used to raise funds for worthy causes, to raise money in aid of men off work through sickness or accident and to raise funds for other community activities. The importance of quoiting was recognized in awards made to individuals for their contributions to quoiting. In every case it brought together different elements of the community. For example, a presentation was made at the Collier Lad Inn, Burradon in 1899 to M. Wardle in recognition of his 56-year contribution to quoiting and the mining community.[54] The celebrations were attended by union leaders, and leaders within the community. Perhaps the most vivid illustration of its importance was the number of poems, written by miner poets, that celebrated the game. For example, Matthew Tate, one of the most prolific of the miner poets, waxed eloquent about the importance of quoits in his life:[55]

> The game of quoits survives us yet,
> It had for me a magic touch
> Above all else that e'er I met,
> Fain would I on its charms descant
> The hobby of my younger years.

The above account of fives, potshare bowling and quoits has treated them as separate entities. While to some degree each had its own coterie of participants and supporters for whom the individual games provided a focal point to life, in reality the boundaries between the three games were blurred and, in fact, they can be collapsed into one category – traditional sports. The centrality of potshare bowling, handball and quoits is illustrated in their inclusion in the poetry of the coalfield. For example, both Matthew Tate and Joseph Skipsey wrote verse lauding their qualities.[56] Additionally they were the most cited games by the 17 miners interviewed in the 1970s.[57] These games, as a totality, reflected the basic elements of the traditional patterns of leisure pursuits. As a group they provide a view of a world very different from that espoused by the dominant figures in the villages, the colliery officials and a particular segment of miners, the adherents of Methodism and "Rational Recreation". The differences between the traditional sportsmen and others were clear and were illustrated, most vividly, in the games they played. It was a rare occurrence when the dominant players of the quoit ground, bowling track and handball alley graced the cricket and football fields or participated in the sport of cycling. They belonged to two different sporting worlds. The traditional games were rooted in traditional locations and forms of competition. All their activities focused on the inns – the series of meetings prior to any challenge match, the social activities after the match, and the innkeepers' promotion of handicaps. Additionally the inns were the site of the competitions, especially in quoits and handball. In fact, it was the development of alternative facilities for quoits in the miner-controlled institutes, in the 1890s, that introduced new elements which eventually heralded the downfall of the traditional sports, in particular the development of clubs and league competition. To put it in its simplest terms, the inns, such as the Astley Arms and Hastings Arms in Seaton Delaval, were a focal point for a series of activities centring on the traditional games. One can imagine that every evening groups of miners would gather at the inns to discuss the past and upcoming matches and handicaps. Frequently they would play matches on the facilities. The games were central to their lives.

There were other elements that differentiated the traditional sports from the new forms of sporting competition. While it is probable that miners played handball and quoits on a regular basis at the various inns around the coalfield, the high profile matches and handicaps were characterized by their irregularity. Unlike football and cricket there were no regularly scheduled competitions. There was either feast or famine. For example the Astley Arms, Seaton Delaval in 1886 promoted only one handicap before September. From thenceforth quoit and fives matches were a weekly occurrence, 10 Saturdays when double-handed fives matches were played and four Saturdays with two quoit handicaps and two matches.[58] The sudden interest in fives was in part due to the opening of a fives alley at the Astley Arms with seating for 700 spectators. The example of 1886 was typical, a sudden flurry of matches and handicaps followed by a total absence of matches. This typified the history of the high profile competitions. In fact, fives disappeared from view after 1902 but was still being played just prior to the war. The example of the Astley Arms was also typical of the inns throughout the coalfield where irregular matches and handicaps were held. At the same time, rabbit coursing, quoits and shooting reflected the fundamental changes that were taking place. First, they witnessed the development of a recognized season, then followed commercialization and centralization.

There were other characteristics of these traditional games that served to differentiate them from the new modern sports. These differences can be illustrated through the history of one man, Henry Pegg, a miner from Seaton Delaval.[59] Henry Pegg first comes into view in May 1886 at the age of 31. He was entered in a quoit handicap at the Astley Arms. During the following 26 years, Henry Pegg played numerous quoit matches and handicaps. His last appearance, of which we are aware, was in September 1912, at the age of 57, in a match at the Astley Arms, Seaton Delaval. However, at the same time as he was a competitor he was also acting in other capacities. From 1886 he frequently was called upon to act as handicapper, referee and, on occasion, stakeholder. The specialization of roles that soon developed in the new sports was atypical in the traditional sports. While Pegg was predominantly a quoiter, he also graced the handball alley, the bowling track and kept pigeons. In fact, he was involved in a novel match over 300 yards which pitted his pigeon against T. Brown, a runner; the pigeon won by five yards![60] Additionally he was called upon to act as handicapper, referee and stakeholder in these sports as well. The fact is that the boundaries between the three games and the roles played within these games were blurred – very different from modern sport.

The history of these traditional sports reveals characteristics of working-class sport that transcend geographical boundaries.[61] In particular, there was one working-class sport played in the working-class villages and towns of Yorkshire, Lancashire, the Midlands, Central Scotland and mining villages throughout England – quoits.[62] Quoits demonstrated certain characteristics that were common wherever it was played: a focus on the inns, the predominance of challenge matches, the slow growth of inter-club competitions, the cyclical nature of its popularity, and its fundamental masculine characteristics of strength, skill and luck.[63] Potshare bowling and handball, on the other hand, demonstrate the power of local traditions. Potshare bowling was only played by the miners of Northumberland and Durham. At the same time it did serve as a point of contact between the two districts. Handball, in East Northumberland, was a purely local sport, while across the River Tyne, in Durham, it was the most popular sport with championship competitions which attracted a large number of spectators on a regular basis. It was played into the 1950s. In Yorkshire and parts of Lancashire knur and spell was widely played.[64] At the same time, there were differences in both the form and chronology of sport. This was particularly the case with quoits. In Central Scotland quoits demonstrated the cyclical nature of the game. The highpoint of quoits popularity was in the 1830s and 1840s. It then went into decline before another spurt of interest in the late 1870s.[65] Additionally, it should be noted, quoits was played within the confines of quoit clubs. In East Northumberland, in the 1890s, it was played within Workmen's Institutes. There was one further thing that held the traditional games together – gambling.

Gambling

Gambling lay at the heart of the challenge matches and handicaps in rabbit coursing, shooting, potshare bowling, quoits and handball. It was so pervasive that, despite ongoing efforts to suppress it, by the public authorities, and persistent condemnations of the evils of gambling by the moralists of the community, it continued to thrive. However, the newspapers, which frequently carried vehement condemnations of the evils of gambling, still provided betting odds for the public. While it appears that gambling was pervasive it is difficult to gain any real insights into the gambling community. Because of its illegality, gambling remained submerged. Strangely "pencillers" frequented sporting events on a regular basis but rarely appear to have been arrested. However, much of the gambling was done in secrecy that rendered it invisible. Nowhere was this invisibility more evident than in the game of pitch and toss. This simple game, in which individuals bet on the result of two coins thrown onto the ground, was, in all probability, the most popular game throughout the century not only in East Northumberland but in mining villages throughout England and in parts of Lancashire.[66]

Although there is fragmentary evidence from earlier in the century it was not until the 1870s that the guardians of public morals addressed the problem of "schooling", as it was called. "Schooling" in Dudley Colliery had become so pervasive in the eyes of the Rev. Blair, Vicar of Killingworth, that he called a public meeting to consider the best ways of putting it down.[67] In 1888 comments were made about the popularity of Sunday gambling in Ashington and Blyth. An interested observer in Blyth commented on the prevalence of "Sunday Schools" on the Blyth links. He commented that "A deep seated liking for gambling at pitch and toss exists in the district; and it is not likely that the vigorous officer (P.C. Metcalf) or the lessons inculcated by our moralists will damp its spirit for some considerable time yet."[68] In Bedlington the comments focused on the public nuisance caused by the "large school of gamblers, either in Coach Road or Walker Terrace" which prevented passers-by from getting past for several minutes at a time.[69] This illegal activity was apparently carried on without interference from the law. The early 1890s witnessed a number of letters drawing attention to the pervasiveness of the game throughout the coalfield and calling upon local councils to take some action against "schooling". These concerns bore fruit in 1895 when Ashington Parish Council wrote to the Clerk of Morpeth County Justices to see what could be done about the "objectionable practice of 'schooling'."[70] The concerns of the 1890s paralleled increased police involvement in attempting to control gambling. As we shall see all these initiatives appeared to have little effect on the pervasiveness of pitch and toss. The evidence for this lies within the court records.

It is through the records that we gain a clearer view of the nature of the game, where it was played, and its popularity.[71] The game was played on public land, on the streets and highways, on the premises of inns and on the recreation grounds that proliferated in the 1890s – in fact, on any spare land.[72] The traditional playgrounds of the miners – the Town Moor, Newbiggin Moor, the links at Blyth and the beaches along the coast – were favorite resorts of the gambling "schools". However, they also played within the villages themselves. In the early 1900s gamblers were convicted for playing on the streets of 11 villages near Blyth. The degree to which the game permeated the villages is illustrated in Hirst in 1910 when 18 prosecutions were instigated for playing on Old House Road, Third, Fourth, Fifth, Sixth and Seventh Avenues, and Poplar Street.[73] There were other locations. For example, the footpaths on Ashington New Moor in the 1890s, between Ashington and Pegswood in 1892, at Bothal Dene in 1894 and near the Bedlington railway viaduct in 1901, were all used by gamblers. Throughout, Sunday was the day for pitch and toss. In the Ashington area various "schools" attained a degree of permanence. There was a "school" named the "Ashington School" that met in Roberts Field. Another school met at Hirst End.[74] This was one of the largest, at which hundreds of pounds were reputed to change hands every Sunday. Yet other schools congregated at North Seaton and Newbiggin Moor.[75] The game was also popular in Coxlodge and New Hartley. It does appear that few villages did not boast a "school". These examples reveal only the tip of the iceberg since they represent those gamblers that were caught. The gamblers took extensive precautions to avoid detection. As early as 1876 comments were made about the difficulty of apprehending the miscreants since "They have their watchers so thickly planted, that it is about impossible for police to go near them."[76] Care was taken to select secluded spots which were ideal for avoiding police. "The groups usually select a nice little hollow piece of ground, and one can nearly always notice their sentinel who occupies a more elevated position, so as to be able to give due warning of the approach of the police, who frequently pay a visit to the links in plain clothes."[77] Later, in 1904, the Sunday "schoolers" visited a remote part of Newbiggin Moor "so that they can 'spot the Bobby' as they term it, as soon as he appears at the edge of the moor."[78]

It is through the prosecutions initiated by the police that we get a glimpse of the extent of pitch and toss, the individuals that took part, the time of year that it was played, the actions taken by the authorities to eliminate gambling and the ultimate lack of success of the authorities. From the first prosecution in 1871 there was a distinct increase in the numbers appearing in court, from 22 prosecutions in 1872 to 307 in 1913.[79] However, this increase was, for the most part, due to the development of the police force and the shift, in the early 1900s, to a concerted attempt by the authorities to erase gambling. This shift, in its turn, is indirect evidence of the extent of the game. Over the years the number of prosecutions for one game of pitch and toss varied between 1 and 27 active players. However, this represented only a small portion of those involved. Often as many as 200 congregated to watch pitch and toss. The participants were drawn from the male population with defendants from as young as 11 to men in their 50s – it truly embraced the whole male population. It knew no season, being played throughout the year. The deep roots and the resilience of the game were reflected in the failure of the authorities to eliminate it. In the 1880s the punishment meted out by the magistrates frequently included one month in prison. This was ameliorated in the 1890s to a fine of 10 shillings or 14 days in gaol. However, there is little evidence that this had any impact on the game. For example, Davison of Morpeth was imprisoned for one month, this was his 23rd offence. In 1910, on several occasions, repeat offenders appeared before the bench – apparently the punishment had little impact upon the playing of the game. Frequently the miners simply did not appear in court to answer the charges. At the outbreak of the war similar comments were being made about the prevalence of pitch and toss as were being made in the 1870s – very little had changed, gambling was central to the lives of a large number of miners.

Conclusion

This chapter has focused on the strength of tradition. What it has done is take us deeper into an understanding of tradition, how it maintained itself and what changed it. What it illustrates is the complexity of traditional practices. In fact, the game of quoits was played in many working class areas of Britain; thus, it could be argued that there was something common to the working class of Britain. That, of course, simplifies the situation since both the form and chronology of quoits were different in different areas such as the villages of East Northumberland and the towns and villages of Central Scotland. It was evident that local traditions were powerful forces and played different roles in working-class life. The differences between potshare bowling and handball in Northumberland and Durham are clear as is the role of knur and spell in Yorkshire, Lancashire and some mining villages in Durham. The whole role they play remains to be investigated. However, it is in the changes that took place that we gain deeper insights into the whole process. These changes were related to external events and, in particular, actions taken by the government, landowners and colliery owners. All across Britain efforts were made to control public thoroughfares, public land, and space in general. One thing that does emerge from this is that the working-class experience, with respect to space, in cities such as London, Birmingham and Manchester, was different to that in the mining villages and smaller towns. It was a reaction to these forces that precipitated internal reactions. This moved the inns and innkeepers to centre stage. They played a critical role throughout working-class areas in the commercialization of sport. In East Northumberland they lay at the heart of the moves to provide facilities for working-class sport. The changing role of the innkeepers had a major impact on sport. While they had always been involved as the site for making matches, the actual organization and running of the events was done by miners. The shift from challenge matches to handicaps fundamentally changed the whole focus and meaning of the games in the lives of the miners – control was taken out of their hands and placed in the hands of entrepreneurs. The whole societal function of the games had changed. At the same time, there was a reaction from within the communities that illustrated, vividly, the strength of working-class desire for independence. The massive explosion of interest in quoits in the 1890s was precipitated by the development of quoits in the Mechanics' Institutes, Workmen's Institutes, and miner-owned and operated social clubs. It is difficult to over-emphasize the importance of this change. By the early 1890s the game was being played on a nightly basis in all the mining villages. The whole focus of social life, for the men, had shifted from the inns to the social clubs and institutes. It truly was a revolution. Finally, there were two other elements that served to differentiate working-class sports from middle-class sports: the cyclical nature of sport and the role of gambling. For example, quoits in both the mining villages and Scotland experienced ups and downs, times when the game was widely played and others when there was a dearth of competitions. There was not a gradual, inexorable development. This was partly due to economic conditions. When miners were fully employed the games prospered and when they were out of work the games disappeared. Related to the uncertainty of life was the popularity of gambling in working-class areas across Britain. Perhaps it was the conditions of life that provided the stimulus for gambling. Miners faced uncertainty, death and injury on a daily basis. Life was to be lived in the present. If you were always living on the brink, what better activity than to take a chance to temporarily improve your conditions or provide for an evening of drinking and fun? Luck was a fundamental element of life.

What, then, can be drawn from this history of traditional sports about the nature of the mining communities? The strong hold the sports still maintained in 1914 attests to the power of tradition. This can, in some ways, be seen as a rejection of the modern sports that were becoming increasingly popular in the villages. Thus, for some groups, these sports provided a link to the past and a rejection of those sports that brought the villages into contact with the outside world through membership in externally controlled organizations and participation with teams and individuals from outside the area. This rejection of the present is reflected most vividly in the lack of success of efforts to form leagues in quoits and, as we shall see later, in billiards. More importantly, these games and the heroes they produced exemplified some of the basic characteristics underlying mining life. The sports, without exception, were individual sports pitting one individual against another. The miners were renowned for their individuality, thus these sports provided a vehicle for them to test themselves against others. While skill was recognized and appreciated there was a great deal of luck and chance that entered into the contests. This was particularly the case in potshare bowling. Of course, this simply reflected life – luck and chance ensured that some survived and some died. Thus their sports were clearly linked into the reality of their lives. There were other elements that suggest that sport was different for the miners than it was for the middle classes. In somewhat of a paradox, winning at any cost was not a part of the dominant ethic. This was reflected in the way in which many champions competed against others with handicaps that tried to make the competition more equal. Frequently champions lost. Another interesting element was the interchangeability of roles – one individual on successive weekends would be player, handicapper, referee or stakeholder. The roles that are so clearly defined in modern sport were not present. Additionally, sport truly was a lifetime pursuit. Even potshare bowling, which required strength and skill, had men who competed for over 30 years. Finally, and perhaps most important, was the centrality of money and gambling to all these sports. Of course, for the miner life was a continual gamble. Every time he went down the pit his life or health was at stake. Thus, it is not surprising that in his sports he gambled and, frequently, lost.

4 The transformation of traditional sports

The last chapter provided a view of the power of tradition and its role in cementing traditional relationships. Cricket, rowing, pigeon flying, athletics and fighting were, in many respects, similar to the traditional sports. However, all changed during the second half of the century. In every case, they acquired some of the characteristics of modern sport while, at the same time, maintaining many of the traditional practices and values. They represent a distinct mixing of the old and the new. Traditional practices and values remained at the heart of these sports while acquiring a degree of organization, a change in the format of competition, and, in some cases, standardized rules. Their histories reflect, most clearly, the tensions between the old and the new.

Cricket

Cricket is not a game that comes immediately to mind when one considers sport in a mining village. This would appear to be borne out by a comment in the *Newcastle Daily Chronicle* in 1901 "That cricket was not well developed in Northumberland."[1] While this was undoubtedly true when the game was compared to football, cricket's roots extended as far back as football's and reflected a combination of the old and the new. Deeply rooted in village traditions, cricket exemplified the past. At the same time, it adopted, albeit reluctantly, elements of the new: the development of regular schedules, the introduction of leagues and the development of enclosed grounds with pavilions. In fact, cricket in East Northumberland fitted in with some general patterns across England, in particular, the growth of clubs in the later third of the century. Between 1854 and 1914, 209 cricket clubs were formed in the area, the vast majority (81 per cent) in the period after 1878.[2] The majority (57 per cent) never saw a first anniversary and another 25 per cent lasted for less than five years. However, there was a group of 26 clubs that were permanent and all demonstrated certain characteristics. All were supported by the dominant figures in the community.

Like all other sports one only gets vague glimpses of cricket in the first half of the century. Henry Shum-Storey of Arcot Hall was a renowned cricketer of the 1820s and there were periodic references to contests between clubs based on the country houses that were built in the 1820s – Arcot Hall, Meldon Hall, Longhirst Hall and

Plate 4.1 A local cricket team in front of the Old Market Inn, Waterloo, *c.*1900. Reprinted with permission from Balmer, R. (ed) *Blyth*, Zaltbommel, Netherlands, 1983

others.[3] These games were rooted in the traditional patterns of social relationships that permeated East Northumberland prior to the arrival of coal-mining. The mixing of the old and the new and the transition from pre-industrial sport forms to modern sport can be seen in Ashington.[4] Ashington Red Rose was formed sometime between 1865 and 1868. During the early years cricket was deeply rooted in the past with the majority of games being played at annual village festivals. For example, in 1875, Red Rose played other teams at the Choppington Brass Band Annual Gala, Ashington Gala, and North Seaton Brass Band Gala. Until the mid-1870s, all Ashington's home games were played on public ground, Newbiggin Moor. In September 1874, Ashington Red Rose moved from Newbiggin Moor to a ground provided by Ashington Coal Company. Red Rose was on its way to permanency. It was at this juncture that other permanent teams appeared.

Between 1878 and 1883 ten permanent clubs were formed; two of them, Seghill (1878) and Backworth Percy (1879), became powerhouses in Northumberland cricket and continue to play at the highest level of cricket to the present day. All the clubs had the support of the dominant figures in the villages. Ministers of the cloth, medical doctors, colliery owners, colliery officials, landowners and school-masters were involved as patrons, administrators, sponsors and players. Five clubs were formed by the landed aristocracy who provided financial support, grounds and frequently graced the cricket field. The first, Longhirst (1878), was formed by the Rev. E. Lawson, owner of Longhirst Hall, who allowed the home games to be played at the hall.[5] In 1887, Sir Matthew White Ridley formed Blagdon Hall Cricket Club, which played in the grounds of Blagdon Park.[6] In 1891, the Samples, agents for the Duke of Portland and residents of Bothal Castle, formed Bothal

Cricket Club with grounds adjacent to Bothal Castle.[7] In 1893 Delaval Hall Cricket Club was formed and played on the south lawn of Delaval Hall with the express permission of Lord Hastings.[8] Finally Delaval Hall Wednesday Cricket Club was formed in 1902 and played at Delaval Hall. Anglican vicars also supported cricket. Seaton Burn Star (1878), Choppington (1892), West Sleekburn (1895) and Cramlington Colliery (1896) were all formed by Anglican ministers who continued to support the clubs both in administrative capacities and on the playing field. Others were promoted by white collar workers. Blyth Wednesday (1894), Ashington Wednesday (1896) and Bedlington Wednesday (1897) were formed by tradesmen to compete on Wednesday afternoons in the league that developed in the 1890s. Additionally, Ashington–Hirst Teachers' Cricket Club was formed in 1902. Even the five clubs formed and run by miners came from certain groups; those associated with the Mechanics' or Working Men's Institute. However, in each instance they were given support by colliery officials or local landowners. More importantly access to a suitable ground brought all these teams within the domain of the land and colliery owners.

Even the impermanent teams were formed by particular groups. In Ashington/Hirst, in the 1890s, 26 clubs were formed, all except two involved teachers, tradesmen, colliery staff, cooperative employees, temperance societies, YMCAs and churches. In the mining villages, without a significant business and professional population, similar patterns emerged. Clubs were formed in conjunction with the Anglican Church (3), Primitive Methodist Chapels (2), coop societies (3), schools (2), temperance societies (5) Mechanics' Institutes (5) and YMCAs (1). Additionally 28 of the clubs were supported by ministers, doctors and colliery officials. Support for cricket came from a clearly defined segment of the community that espoused the values of self-help and education. Unlike the deeply rooted popular sports of bowling, handball and quoits, cricket found fertile ground amongst the very groups that eschewed the traditional sports that focused on the public houses.

The involvement of particular segments of society in cricket can, in part, be explained by a basic reality of cricket, the need for a ground to play on. This brought cricketers within the domain of the landowners, colliery owners and colliery officials. It was the need for pitches that was the reason for the failure of so many of the impermanent teams. As the century progressed, the shift from playing at the annual festivals to the development of regular schedules placed a premium on the acquisition of a permanent ground and, thus, brought them within the purview of the landowners. By the 1890s an enclosed ground with a pavilion became a basic requirement for the success of the permanent clubs. From the earliest days, cricket clubs depended upon the largesse of the landowners. Ashington Red Rose claimed their success in 1874 was due to the actions of the Ashington Coal Company in providing a field for them to play on, and Seghill, a club destined to become one of the most successful, owed its start to a field granted by Mr Joseph Laycock, the owner of Seghill Colliery.[9] During the 1880s and early 1890s Longhirst Hall, Blagdon Hall, Bothal Castle and Delaval Hall became the sites of grounds provided at some considerable expense by their owners. From the early 1880s, the most prominent clubs embarked upon efforts to improve their grounds by levelling and re-turfing, providing a pavilion and more extensive facilities, and enclosing the grounds to keep out unwelcome spectators . In some cases, they charged admission to the games. New Hartley Hastings, Blyth, Backworth Percy, Newsham and Seghill were heavily dependent upon donations from the land and colliery owners.[10] Others were given support from figures more closely linked to the community, in particular local farmers. Finally, with the development of community recreation grounds, several clubs moved onto these community controlled grounds – East Holywell and Seghill Recreation Grounds were used for cricket in the early 1900s.

The basic question that remains to be answered is to what degree did miners participate in cricket? The fact that it had some hold in the communities was demonstrated by the presence of seniors and juniors at the West Sleekburn Colliery Brass Band Gala in June 1876.[11] This was followed, in July, by the West Sleekburn Cricket Club Gala which included matches between the cricket club and an all-comers team and another match involving two schoolboy elevens. On June 4, 1881, Seghill played Longhirst; ten of the Seghill players and six of the Longhirst team were miners.[12] They were joined by a colliery weighman, a railway engineer and a railway fireman. Later in the summer Longhirst played the newly formed Hartley Hastings, which included seven miners plus Dr McCall and W. Beck the colliery engineer.[13] Later during the strike of 1887 the success of Ashington Red Rose was attributed to the fact that the Ashington men had more time for practice since they were on strike![14] It was in the summer of 1883 that Newsham Jolly Lads played a series of games against teams from Bates Cottages, Backworth, Blyth and New Delaval.[15] After a game in Holywell a crowd "of over 200 of the Holywell people stoned the Newsham 'Jolly Lads' cricket club when they were leaving the field." It transpired that that the Bates Club comprised schoolboys and boys who worked down the pit. Seghill, one of the most successful teams in the district, in 1896, included Thomas Mulby (27), George Taylor (37), and John Landreth (47), all miners working at Seghill Colliery.[16] Later, in 1899, the members of Holywell Mechanics' Institute formed a cricket club; the president, M. Davey, was a 33 year old miner.[17] In fact, it was through the Mechanics' Institutes, closely associated with the colliery owners, that miner controlled clubs emerged – Bebside Juniors (1884) and Dudley Weetslade (1897). These, in conjunction with two clubs formed by social clubs in Ashington, are the only clubs that can unequivocally be linked to miners, although it is possible and indeed probable that teams like Dinnington War Cry and Boca Chica Magic (1889) comprised miners.

In fact, cricket provides concrete evidence of the persistence of traditional practices where different social groups played together or, at least, were involved in the same sport together. The relationship can be demonstrated most vividly in the case of Hartley Hastings, one of the most successful teams in East Northumberland. From its formation in 1881, Hartley Hastings brought together the social elite of New Hartley, Seaton Delaval and Seghill. At the same time, miners were present on the playing field.[18] In a match against Longhirst, seven of the players were miners. They were joined by Dr Howard. During the ensuing years the social elite were regular participants – Rev. S.C. Whalley (1887); Dr Wallace (1889); and E. Thompson, Lord Hastings' estate agent, played for several years in the 1890s and was "best bat" in 1896. The social elite played an active role in the affairs of the club. Throughout the 1890s, Lord Hastings, Lord Carrington and Mr R.O. Lamb, colliery owner, were patrons who attended meetings and provided financial and moral support to the club. On several occasions, Lord Hastings acted as president of the club. At the annual meeting of the club, held at the Astley Arms Inn on February 9, 1892, it was commented that "Many of the patrons of the club [were] gentlemen of rank and influence."[19] These men were drawn from outside the immediate vicinity with A.E. Burdon, of Hartford Hall, serving as chairman at the annual meeting in 1897. The vicars and curates of Seghill and Seaton Delaval were actively involved as were the doctors of New Hartley, Seghill and Seaton Delaval. It was, perhaps, the social standing of the patrons that ensured that the Hartley Hastings cricket ground was recognized as one of the best in the county. At the same time, since financial support was frequently solicited from the miners of New Hartley by members of the teams going around the village with a box, it is likely that the miners continued to play on the team. Thus Hartley Hastings Cricket Club carried on the traditions of Olde England where Lord and servant played on the same pitch. That Hartley Hastings was not atypical was reflected in the composition of Ashington Cricket Club in the early 1900s. Players included two miners, two deputies, a joiner, a blacksmith, a butcher, a doctor, two schoolmasters, a bank clerk and the mine manager.[20]

Cricket achieved a degree of permanence in the 1880s as the number of permanent clubs increased. However, yearly schedules were still characterized by their variability from year to year. This changed to some degree in the 1890s when Allotment Rising Star joined the Tyneside Junior Cricket League. Unlike Lancashire, Yorkshire and the Midlands, league cricket with its regular schedule of games and its implicit increased emphasis on victory was never adopted with much fervour by the cricket community.[21] However, leagues did come to the district and were important in regularizing the schedules. Allotment's brief flirtation with league cricket was followed in 1897 by the first attempt to form an intra-district league.[22] The Blyth and District League was formed in 1897 by the five leading clubs in the area: Seghill, Backworth, Ashington, Blyth and Morpeth. It disbanded on January 23, 1899 as a result of internal squabbles over the state of the grounds and the desirability of league cricket. Subsequent efforts to promote internal leagues suffered the same fate; the East Northumberland League (1904–05), Blyth and District League (1906–07) and the Ponteland and District League (1905–07) all collapsed after a brief existence. The only successful initiatives were those that involved East Northumberland teams with those outside the area. The Northern Wednesday League (1901–10) brought teams from Blyth, Bedlington, Ashington and Delaval Hall into competition with 16 teams from Northumberland and Durham. However, the most important league was the Tyneside Senior League, which was formed in 1903. Morpeth (1905–10), Backworth Percy (1906–), Seghill (1909–) and Blyth (1910–) were the pre-eminent clubs in the area and the only ones invited to join the league. Incidentally these were the only teams that hired professionals. The advent of league cricket brought some stability to the summer schedule. This was paralleled in the other clubs by the development of more regular schedules. After 1894 not many clubs' schedules listed fewer than 17 games a season. Thus, Saturday afternoons became the focal point of cricket fans throughout the summer weekends.

One final question remains to be answered: how does cricket in East Northumberland relate to the game in the rest of the country?[23] In fact, it was an amalgam of the old and the new, of the south and the north, of country house, village, and league cricket. At the same time, it demonstrated similar patterns of development, especially in the role of the dominant groups, the churches and chapels, and the institutions of "Rational Recreation" in promoting the game. The differences lay in the timing of the changes and the depth of development. Throughout Britain cricket witnessed a veritable explosion of interest during the last third of the century. In the towns and villages of East Northumberland it was more modest and occurred later, during the 1890s. Additionally league cricket, the defining feature of cricket in the industrial regions of the Midlands, Lancashire and Yorkshire, which developed between 1873 and 1892, arrived in East Northumberland with any degree of permanence only in 1903. League cricket was beset by difficulties even with clubs at the highest level. Only Backworth and Seghill, two quintessential mining villages, showed a long-term commitment to league play. The other three, Blyth, Ashington and Morpeth, were reluctant participants but eventually succumbed to the attraction of league cricket. It was no accident that these three clubs considered themselves amateur. They had more of an affinity with the clubs located on the estates of the landowners who continued to play exhibition games. Their identification with the traditions of cricket exemplified in amateur cricket was demonstrated graphically at the annual supper of the Blyth Cricket Club in 1912 when Mr J.W. Parsons stated:

> It was not, however, the winning of games that constituted the chief charm of cricket, but the discipline, the esprit de corps, the self-respect, and respect for others that was the attraction of cricket, and by which its moral, as well as physical lessons, were inculcated.[24]

Rowing

In several respects, the history of rowing demonstrated some of the same characteristics as cricket: the focus on limited locations, the origins rooted in pre-industrial social relations, and changes as a result of its brush with the modern world. However, these similarities were overshadowed by the changes that precipitated the sport into the modern world. Traditions were destroyed, although, as we shall see, not totally obliterated, and rowing emerged as a sport where the class lines were clearly recognized and enforced. By the turn of the century rowing, which in the 1860s was the first truly popular spectator sport, had become the preserve of a small coterie of participants competing on an irregular basis in front of small crowds.

Rowing is not a sport that one naturally associates with coal-miners, yet it is possible to argue that it was the first sport to attract widespread support throughout the coalfield.[25] However, like cricket the sport took root in limited locations, 23 villages on the banks of the Rivers Blyth and Wansbeck. In fact it was more localized, with 70 per cent of all rowers coming from Blyth, Cowpen and Bebside on the River Blyth and North Seaton and Cambois on the River Wansbeck. At the same time, the high profile races involving the top rowers from across the world attracted spectators from all the mining villages. This was particularly the case with the competitions held on the River Tyne during the heyday of the sport in the 1860s and 1870s. It was an immensely popular spectator sport with thousands of spectators trekking to the banks of the Tyne between the High Level Bridge and Scotswood to watch the highly publicized professional competitions for stakes as high as £1,000. Perhaps even more illustrative of the tremendous popularity of rowing amongst the working class was the number of mourners attending the funerals, in 1868 and 1870, of two of the greatest of all Tyneside rowers, Robert Chambers and Harry Clasper. Their funerals attracted over 200,000 mourners, numbers fit for a king, and kings they were on the banks of the Tyne.[26] Many of these mourners came from the pit villages of East Northumberland where rowing, both as a participant and spectator sport, had its pockets of support.

Between 1858 and 1878 there were a number of regattas held on the Rivers Blyth and Wansbeck. (See Table 4.1). Most were marked by their impermanence. The one with the greatest degree of continuity was the Great Northern Regatta which was held at the mouth of the River Wansbeck on a relatively consistent basis from 1860 until its demise in 1874. However, there were certain characteristics that were common to all regattas. First, they were strongly supported by the dominant groups in society. Unlike many of the other sports it was not only the colliery managers that encouraged rowing but the owners themselves. Patronage and pecuniary support for the Great Northern Regatta were provided by several of the

Table 4.1 Regattas on the Rivers Blyth and Wansbeck, 1858–78[i]

	8	9	60	1	2	3	4	5	6	7	8	9	70	1	2	3	4	5	6	7	8
River Blyth																					
Blyth	X	X	X					X		X	X					X	X		X		
Bebside & Sleekburn					X	X	X														
Bebside & Cowpen								X	X	X	X										
Blyth & Cowpen										X	X	X	X	X						X	
River Wansbeck																					
Great Northern			X	X	X	X	X	X		X	X	X		X		X					
West Sleekburn																	X				

i Data for the Table from NDC, 1858–78; MH, 1858–78.

East Northumberland mine owners. Additionally it was "liberally patronized by William Watson, Esq. of North Seaton Hall, and the proprietor of the royalty of the colliery."[27] The first regatta was sponsored in 1860, the same year that North Seaton Colliery was opened. The managerial staff of the colliery played an active role in the proceedings, acting as judges. However, the regatta was organized by a committee of miners from Cambois and North Seaton Collieries. This truly was a communal occasion with festivities continuing into late in the evening.

The Great Northern Regatta also serves to illustrate the other characteristics that were common to these events: the programme of events and the activities surrounding the regatta. At the inaugural regatta in 1860 there were eight events: three sailing races, four oared races and one coble race. In the following year the eight events attracted 50 competitors. By 1862 the event attracted 82 competitors who rowed cobles, keels, jollys and gigs the three-quarters of a mile up to the turning point and back. Boats such as the "Morning Star", "Little Infant", "Hawk", and "Pride of the North" vied for the money prizes that were awarded to the winners.[28] For the Fifth Annual Regatta in 1864 the programme had been expanded to 14 races. More importantly there were events for the legitimate racing craft, the skiffs. These events began to attract the top professionals to the regatta. This was, in fact, the high point of the Great Northern Regatta. By 1867 there were only five races, one of which was a six mile sailing race out to sea. The organizers had also introduced a 400 yard swimming race which attracted 13 competitors. The inception of skiff racing led paradoxically to the demise of the regatta. The 1860s saw the growth in popularity of professional rowing and with it an increase in purse size. The committee of miners recognized the lack of publicity and the need for higher prize money. However they were unable to offer sufficient prize money to attract the top professionals and, thus, interest in the regatta declined and it disappeared after 1874 – in part, a victim of its own success.

The annual regattas were communal celebrations providing a variety of activities and experiences where the rowing events were handicaps that involved several crews. More central to the sport of rowing were the matches between two rowers that were rooted in the traditional forms of competition, the challenge matches. Although all challenges were for money stakes, there was a distinct difference between the purely local competitions and those involving men who made a partial living from rowing and whose careers stretched over several years. At the purely local level, rowing challenges followed no regular pattern although there was a distinct decrease in the 1890s. There were infrequent but persistent challenges between different groups of miners from Cambois, Cowpen, Bebside, North Seaton and other collieries. They provided an ongoing but weak link to the rowing scene.

For a small group of miners, professional rowing provided an avenue out of the mines. For the majority it was a temporary respite, but for others it was a vehicle to escape the mines on a more permanent basis. The opening up of these opportunities paralleled the introduction of skiff racing to the northern rivers in 1864. One of the earliest of these miners who availed himself of the opportunity to get out of the mine, was James Boyd, a 24-year-old Blyth miner.[29] The history of James Boyd's rowing career (1864–74) encapsulates the whole world of professional rowing, the complex relationships surrounding any match, the money involved, and the opportunities to escape the mine. The arrival of skiff racing on the northern rivers led to the formation of a number of clubs, the Blyth Club, the Wansbeck Club, the Fawcett Club and the Matthew Potts Ridley Arms Club. These clubs were one boat clubs of professional rowers focusing upon particular inns. A series of races for stakes were organized between the different clubs. James Boyd was associated, in the early years, with the Blyth Club. During the period 1864 to 1874, he participated in a variety of challenge matches for stakes of up to £100 and in a variety of handicaps. Each of his challenge matches involved a set of negotiations in which Boyd and his backers met with his opponent and his backers at an agreed upon inn. These ended in the signing of articles of agreement which laid down the conditions of the race and the meetings, at inns, at which the stakes were to be deposited. Boyd then entered into a period of training which varied from five weeks to two months. The training was centred on an inn adjacent to the river. In the early years, Boyd trained under the tutelage of the great Harry Clasper. The training sessions frequently attracted large numbers of spectators to the river and the sponsoring inn. By 1867 Boyd had been so successful that instead of the races being a break from his employment in the mines he was able to make a break and become a boat builder. As his career began to decline in 1873 with two consecutive losses, Boyd moved into the promotion of rowing and began promoting handicaps on the River Blyth. James Boyd's career paralleled the great era of rowing on the Rivers Blyth and Wansbeck. From 1864 until approximately 1880, challenge matches and handicaps were held, on an annual basis, on the two rivers. However, the extent of the racing should not be over-emphasized, as 1874 demonstrates: three matches and one handicap.

There was a small group of rowers, all local miners, who emerged during the 1870s, whose fame spread beyond the boundaries of the coalfield and onto the premier rivers of the rowing world, the Tyne and the Thames. William Elliott (1874–80), a Pegswood miner, epitomizes these professionals and their importance within East Northumberland.[30] Elliott first came into view competing in a handicap held on the River Blyth in October 1874. During the next three years, he won a series of minor races on the local rivers. It was not until March 6, 1877 that the 27-year-old competed on the championship course on the Tyne. After a hotly contested match with William Nicholson he picked up the £200 in stake money. His status within the community was demonstrated the following week when a gathering of local citizens and fellow oarsmen gathered at the Theatre Royal, Blyth to present him with a gold watch.[31] This race precipitated Elliott into his most successful and busy year, 1878. After winning the £100 stake against a leading Tyne rower, James Bagnall, early in January, Elliott and his backers moved to arrange a return match with William Nicholson. The four deposits were made at four different inns in Newcastle. Elliott started training under the watchful eyes of James Taylor and a new skiff was specially constructed for the match. Elliott won the £200 stake. From there Elliott moved south to the River Thames where he successfully raced against a Putney professional for £200. He reached his peak on the June 8 when he raced J. Higgins of Shadwell for £400 and the Championship of England. He lost. In six months Elliott had competed for £900 in stake money. The defeat did not lessen his popularity amongst the citizens of Blyth. On his return on September 28 an "immense" crowd greeted him at the railway station. He was carried through the town in a procession led by Cowpen Colliery Brass Band.[32] Elliott continued to compete at the highest level until the early 1880s. In fact, the highpoint of professional rowing had passed and the sport began to recede into the background. The number of challenge matches decreased and rowing became a rarity. However, histories of a number of the local professional rowers of the 1870s illustrate the tenacity of tradition and a different approach to competition.

The persistence of the old ways and the different meaning of competition was revealed in the history of Lewis Lewis, who graced the Blyth and the Wansbeck for an incredible 46 years (1866–1912). From his entrance into the "aquatic" fraternity in 1866 until his match for a £100 stake against his old rival Anty Scott, on the River Blyth in 1912, Lewis Lewis competed in approximately 40 contests on the two rivers.[33] Contests provided welcome breaks, of two to three months, from the hardships of the mine but never provided a livelihood. However, Lewis's prowess in rowing eventually allowed him to get out of the mine in 1910, when he took over the Black Bull Hotel in Blyth. Additionally his history revealed the spasmodic nature of competitions – there were periods of feast and famine, it was never as structured and regular as sport eventually became. Unlike many sports of today, rowing was a lifetime sport and even as late as 1910 "Lewey . . . now grey and grizzled, but still very straight and vigorous" competed against his old rival Anty Scott for a stake of £100 – appropriately the race ended in a dead heat.[34] What "Lewey" and his rivals represented was the continuation of the old ways of competing and a different meaning for sport. "Lewey" was rooted in the past where individuals participated in many sports. Our last view of "Lewey" is on December 26, 1913 on Newbiggin Moor where he won a potshare bowling match for £10 a side![35] When he and his colleagues disappeared the old had given way to the new.

As professional rowing declined on the Tyne so too did it decline on the two northern rivers. At the same time it maintained a presence, not so much in terms of the professional rowers but involving local rowers and spectators. These initiatives were due to the actions of a number of innkeepers with hostelries on the banks of the rivers. While the inns had always been the centres of professional rowing, there is no evidence that they encouraged the sport on a broader base. This changed, in the 1880s, when they promoted handicaps and regattas. In both cases the object was to attract spectators and, in the case of the regattas, to utilize boats owned by the innkeepers. Between 1884 and 1896 twenty regattas, sponsored by innkeepers, were held on the Rivers Blyth and Wansbeck. The regattas promoted during the 1880s were significantly different from those of the 1860s in that they were promoted as commercial enterprises specifically designed to make money for the innkeepers. In September 1884 George Trotter of the Rose and Crown, Bedlington Furnace promoted a regatta at Bedlington Furnace on the River Blyth. This regatta, which came to be known as the Northern Collieries Regatta, was specifically designed to attract miners from seven pits. His initiative met with success as crowds estimated at over 4,000 attended the regatta. George Trotter promoted the Northern Collieries Regatta, on a relatively frequent basis, between 1884 and 1896. Similar initiatives were taken by innkeepers of Newbiggin (1885–89) and Sheepwash (1888–90) with the specific purpose of "providing entertainment for the holiday seekers from the adjacent colliery villages."[36] All the events were handicaps in order to attract the maximum number of participants. Additionally only one event for sculls was held, the traditional work-craft being the favourite boats – cobles, keels, pleasure boats. There were other events attracting different segments of the community – a fishergirls' race at Newbiggin in 1887 and a "swimming football competition" at Bedlington Furnace in the same year.[37] This was followed, in 1890, by a water polo competition at Sheepwash which attracted teams from Ashington, New Winning and Cambois. By 1891, the regattas at Sheepwash and Newbiggin had disappeared while the Northern Collieries limped on until 1896. The days of the regattas in East Northumberland were over.

The disappearance of regattas after 1890 was paralleled by the gradual decline in challenge matches. In 1893 on the occasion of a challenge match on the River Blyth for stakes of £120, it was lamented that this was the first major race on the river in six years. The light of rowing was kept flickering by those with a vested interest in its continuance, the innkeepers with riverside hostelries who had a supply of racing skiffs. One of the innkeepers, who actively promoted racing during his tenure at the Seven Stars Inn in North Blyth, was George Hudson. In 1893, he attempted to attract sponsors to form a North Blyth Rowing Club. This collapsed after less than a year. He also promoted annual handicaps in 1893 and 1894. His successor, George Coulthard, continued the practice in 1898. However, they were rarities on the river. Rowing was saved, in 1899, by the formation of the Blyth Rowing Club.

In September, 1899, Thomas Queen of the Golden Fleece Inn, Cowpen Square called a meeting at his inn to consider the formation of a rowing club.[38] He was successful and the Blyth Rowing Club was formed with headquarters at the inn. This club was different from any previously created clubs and brought rowing into the world of modern sport. The "clubs" that had flitted briefly across the rowing scene in the 1860s and 1870s comprised a group of men who invested in a club to support a limited number of professional rowers. Blyth Rowing Club was formed by a group of rowers with the innkeeper providing the racing boats for practice and competition. The Blyth Rowing Club became the focal point of competition on the river and provided the organizational structure for rowers to compete on the Tyne and Wear. From 1900 the high point of rowing on the Blyth was the Isaac Tucker Challenge Trophy for single sculls. It was sponsored by the Blyth Rowing Club over a half mile course. Held on an annual basis until the war, the Tucker Trophy attracted rowers from the Blyth vicinity.[39] At the first, in 1900, 36 competitors from villages on the River Blyth competed with six competitors from the River Wansbeck. In fact, this was the high point of the Tucker Trophy as numbers gradually declined until in 1910 only 24 entered and 14 actually competed. At the same time the day's racing still attracted attention. In 1910, the

Plate 4.2 Blyth Rowing Club *c.*1900.
Reprinted with permission from Balmer, R. (ed) *Blyth*, Zaltbommel, Netherlands, 1983

Morpeth Herald commented that it was like the old days. "From the windows in the miners' cottages at the 'Square' which run along the river, the ladies were to be seen – and heard too – encouraging their favourite oarsmen."[40] While the Tucker Challenge Trophy was the centre piece of rowing on the River Blyth the club was also involved in promoting and participating in the sport in the North East. In particular, in February 1906, they were the leaders in forming a Professional Boat Race Association whose primary mandate was the formation of a professional Four-Oared League.[41] They joined with four clubs based on the River Tyne in this enterprise. It failed. More successful was their entry into the district rowing scene through their participation from 1905 in the Tyne Christmas Handicap, which had been in operation since 1875, and in the prestigious amateur Durham Regatta held on the River Wear since 1814. At the Durham Regatta the Blyth crews competed in the professional four-oared race – the only professional race in a programme of 10 races. Interestingly, the amateur events attracted only two or three entrants while the professional race attracted up to 32 crews – a significant difference.[42] By 1912, the Blyth Club along with two recently formed clubs on the Wansbeck, were important competitors in the Durham Regatta.

The development of the Wansbeck Club and the North Seaton Rowing Club was different from that of the Blyth Club. The formation of the North Seaton Club was more deeply rooted in traditional relationships, in particular the support of the colliery officials. In 1896, a group of 25 miners at North Seaton Colliery attempted to form a club and approached Mr J. Nicholson, manager of the colliery, to be president of the club.[43] The club, liberally supported by the colliery officials and some local gentlemen, held its first handicap on September 19. However, after two or three years it disappeared from the scene. A North Seaton Club reappeared in 1910. Again, it was given financial support by the colliery owners and, in 1913, a new boathouse was opened at North Seaton. The boathouse had been constructed by workmen from the colliery with the active support of Mr Manderson, the colliery manager. The managerial staff at the colliery played an active role in the affairs of the club with Manderson becoming the president and J. Willis, the resident engineer, the vice president. The paternalism of the North Seaton Coal Company, so evident in the development of the Great Northern Regatta in 1860, remained until the First World War. However, the rowers from North Seaton and the Wansbeck Club at Sleekburn joined their brethren from the Blyth Club in competing in the professional contests on the Tyne and Wear. Although claims were made for amateur status they faced the class-based definition of the Amateur Rowing Association – they literally remained beyond the pale.

Rowing in East Northumberland was nearly totally divorced from what was going on in the rest of Britain and even on the neighbouring River Tyne. The patterns of development on other rivers, the growth of amateur regattas and clubs between 1830 and 1880, the commercialization of regattas, and the ongoing battle between amateurism and professionalism found little resonance on the Northumbrian rivers.[44] Only when the small number of professional rowers ventured out onto the Tyne and Wear did East Northumberland rowing touch the mainstream of rowing. They literally lived in a backwater protected against the battles that swirled around rowing in the rest of Britain. For the most part rowing was inward looking with regattas, handicaps, and challenge matches involving local rowers. At the same time, there were tentative, and eventually unsuccessful, efforts to commercialize rowing. In the 1900s there were two efforts to develop rowing clubs along the lines of the clubs that developed in the rest of the country. With the support of colliery officials attempts were made to foster "amateur" clubs. They were thwarted by the class-based Amateur Rowing Association and thus were forced to compete in professional competitions. At the same time rowing maintained a small presence on the rivers and never disappeared totally.

Pigeon racing

The power of tradition, the social relationships that characterized the mining villages, and the impact of the changes of the second half of the century are the themes that provide a focus to the chapter. The history of pigeon racing provides yet another view of these themes and, in particular, the nearly complete separation of the sport into two distinct forms that gained their support from different parts of the community. One, short distance racing, lay within the traditions of mining society and amongst the miners themselves. The other, long distance racing, epitomized the characteristics of modern organized sport and its base of support was more broadly based.[45]

Like many of the miners' traditional sports pigeon racing competitions remained rooted in individual challenges based within the villages. Of 71 challenge matches in 1890 only three involved inter-village challenges and, even then, with villages in very close proximity.[46] These local matches, which rarely involved distances of above 10 miles, remained the most pervasive form of competition throughout. The exact extent of these local competitions remains, for the most part, hidden from view. At the same time, during the second half of the century, pigeon racing moved within the orbit of modern organized sport. Its history reflects trends that should be familiar: the take off phase in the 1870s, the development of formally constituted clubs, the chronic lack of stability, the centrality of the inns and innkeepers in the promotion of the sport, and the importance of the more permanent clubs in the creation of an organizational infrastructure.

The first homing club, the Great North of England Columbian Society, was formed in Choppington in 1877 to promote racing contests between its members. Later in the same year, in Bedlington, which became one of the strongholds of the sport, the Bedlington Homing Club held its first competition. The majority of clubs had a brief existence. Of 129 identifiable clubs only 14 (11 per cent) lasted for more than four years. At the same time there was a degree of continuity – homing clubs were ever present. For example, in the small village of Dudley there were only two years between 1880 and 1900 when there was not a club associated with either the Clayton Arms Inn or the Dudley Hotel.[47] Nearly without exception these clubs were short distance racing clubs. What this meant was that small groups of aficionados, usually around 20, met at the club headquarters, the local inn, with their birds. A club member would take the birds to the required distance, frequently one or two miles away, and release them.[48] The competitions sponsored by the clubs were one form of competition. Just as popular were handicaps and sweeps sponsored by the innkeepers, frequently for members of particular clubs but also on an open basis. These short distance races were the bread and butter of the village homing clubs and the competitions of less than 10 miles accounted for 84 per cent of all competitions. Even in the early 1900s, in Seaton Delaval and New Hartley, mile or half mile races were held every Baff Saturday. These races were over measured miles or half miles from the cree. The birds were trained every night to fly back to the cree from any point of the compass. On the day of the competition each man threw 5 shillings into a pot. At that time all stop watches were synchronized. The birds and one stop watch were taken to the release point while the owners returned to their crees with another stop watch. All the birds were then released. Events of this nature were held in many villages – short distance racing remained, throughout, the dominant form of competition amongst miners.[49] For the pigeon fanciers, caring for the pigeons would take up all their spare time as Jack Dorgan's uncle George's 100 pigeons attest.[50] While the majority of clubs disappeared after a year or two there were a small number that became permanent fixtures on the pigeon racing scene and these were, in several respects, different from the other clubs; in particular they focused on long distance racing.

The history of long distance pigeon racing brings pigeon flying within the realm of modern sport with regularly scheduled competitions, the development of district, regional and nationwide organizations, and the codification of rules and regulations. Although the development of long distance flying occurred in the

1890s, birds had been sent to Edinburgh, Glasgow, Doncaster and York during the 1880s. Of course, the problem is when does short distance racing become long distance? It would appear from later developments that any release point within 10 miles was regarded as short distance. The first of the permanent long distance clubs was the Burradon and West Moor Homing Society (1891).[51] By 1911, there were 21 Northumberland clubs affiliated with the Northern Federation – a small but significant group. It was this group of clubs that led in the development of long distance racing.

The first step in the development of the sport took place in the early 1890s. This was the formation of district organizations that brought a number of clubs together. They, in turn, affiliated with regional and national organizations. At this juncture the local fanciers moved into the mainstream of sport and thus contact with groups from outside the mining districts. The initial move to develop a local organization occurred in Newcastle on July 21, 1892, when the Newcastle-Upon-Tyne Homing Society was formed.[52] This club attracted some of the leading fanciers from Bedlington, Cramlington and Backworth, the majority of whom were businessmen, doctors, innkeepers and colliery officials. However, as with other sports, the idea of a Newcastle dominated organization controlling all Northumberland sport did not go down well within the coalfield. On October 22, 1900 at a meeting at the Blagdon Arms Hotel, Cramlington, the Northumberland Federation was formed.[53] Seven local clubs had been called to the meeting by J.W. Parkin of Cramlington, the dominant figure in pigeon racing prior to the First World War. By 1910, the Federation had grown to 24 clubs with 291 individual members. It organized long distance racing within the district and in so doing affiliated with the Up-North Combine and the National Homing Union. The local administrators became involved, at various levels, in the national scene. Within the coalfield the federation acted to develop a racing calendar which lasted from May until August. This entailed a combination of old and young bird races. The old bird season consisted of five races between May and early July with the first usually being from Doncaster and the last from somewhere on the continent, frequently Rennes, France. The young bird season followed with four races from locations such as Doncaster, Nottingham, Rugby and Oxford. In the off-season, the long distance clubs held annual shows of racing pigeons. The first was the Bedlington North of England Society in 1897. They were followed by Morpeth (1899), Ashington (1900), New Hirst (1901) and others. The most important was the inauguration of the Northern Federation Show in 1905 which was held in Morpeth Town Hall.[54] By 1907 this attracted the largest number of racing birds of any show, 500. This brought the colliery clubs within the orbit of other clubs since breeders brought pigeons from as far away as Yorkshire and Lancashire. Additionally, judges were provided by the National Homing Union from various places in England. The pigeon fanciers of Northumberland were brought into contact with fanciers from all over England.

Long distance pigeon racing required a considerable degree of organization. The logistics of getting the birds to the release point in Rennes in France was considerable. The arrangements for timing, keeping records, and the organization of shows required considerable organizational skills. By 1907, the Northern Federation had delegated the responsibility for transporting the 23,982 birds that were sent that year to the Up-North Combine.[55] The Up-North Combine was formed to negotiate with the various railway companies on behalf of the pigeon racers. Pigeon racing had become an organized sport requiring finances and organizational skills. These skills were provided by doctors, businessmen, and innkeepers. However, the most important of these officials was J.W. Parkin – President of the Northern Federation on several occasions, Northern representative at the Up-North Combine, and a member of the Council of the National Homing Union. Parkin was a deputy overman at West Cramlington Colliery – the lowest administrative position down the mine.

On the surface, at least, the history of pigeon racing revealed two different elements of life in the villages. The two forms, short and long distance racing, were very different in all respects. Short distance racing was located within the traditions of the villages and maintained a robust presence until the First World War. Challenge matches, group contests, and events at communal galas, with the associated money prizes and gambling, were still in evidence at the outbreak of the war. Long distance racing, on the other hand, presaged the future – association with organizations and people outside the coalfield which led to the slow erosion of the isolation of the villages. The involvement of the miners in these clubs is less certain. The clubs were supported by businessmen, doctors and publicans, although the role of J.W. Parkin indicates that some working miners were involved. If this was the case it points to a less rigidly stratified society than one would expect.

The history of pigeon flying in Britain remains to be written. However, it is certain that it was popular in the mining villages of England and, like East Northumberland, "during the second half of the century the sport became more competitive, more respectable and more highly organized."[56] Additionally, it does appear that short distance flying was the most popular in mining communities and that long distance flying was not common to all mining villages. Long distance flying certainly was more broadly based, in terms of participants.

Athletics

Writing in 1828, MacKenzie stated, of the miners, that one of their most popular pastimes was foot racing. The competitions took place on highways and public space and in the annual festivities called "hoppings".[57] In some respects things did not change during the course of the century; miners still challenged each other to demonstrate who was best and still competed in foot racing within the communal festivities. The two streams, pedestrianism and the annual festivities, eventually joined in the 1890s.

Pitmen were competitive by nature, always challenging others to tests of speed and endurance. Challenge matches for side stakes became the dominant form of competition. By the 1840s they had become so popular that the magistrates at Morpeth imposed a 40 shilling fine on anyone convicted of running races on a public highway.[58] During the 1850s the number of competitions increased, with matches being held on Blyth Links, Morpeth Common, the Town Moor, old waggonways and highways leading to Blyth. It was during the late 1850s that the process of commercialization started with the opening of commercial running grounds in Newcastle. Within the coalfield, the growing popularity of foot racing led, in 1862, to seven innkeepers within the colliery villages promoting handicaps on the roads next to their hostelries. In each case the innkeeper offered a money prize in addition to the accumulated entry fees. The combination of inns, money prizes, entry fees and handicaps typified pedestrianism.[59] The inns acted as the focal point for making matches, signing articles of agreement, settling of accounts, sponsors of the races and, frequently, as the location of the contest. However, it was the police and magistrates of Blyth who precipitated full-blown commercialization within the coalfield. In the early 1870s, as a direct result of the increase in contests on the roads leading to Blyth, the police prosecuted a number of miners who used the road for competition. This was the effective end of road racing. The unintended consequence of these moves was to force these matches into enclosed grounds – and full commercialization.

The early 1870s, amongst the most prosperous years of the nineteenth century, precipitated two events that demonstrated, graphically, the increased popularity of pedestrianism, In 1872, a group of Morpeth businessmen inaugurated what was to become the most successful and prestigious of all athletic promotions, the Morpeth Olympic Games.[60] The games, which continued until the late 1950s, offered a varied programme of athletic events, paid the highest prize money, attracted competitors from the North of England and Scotland, and regularly attracted over 10,000 spectators. From the outset it was based on one basic premise – it must pay its own way. It was, above all, a commercial enterprise.

On May 17, 1873 John Dowson, a Bedlington butcher, opened the Bedlington West End Running Grounds, the first commercial sports ground in East Northumberland. This was followed, in 1874, by the opening of Bedlington East End Running Grounds. The offerings at Bedlington West End Running Grounds illustrate the nature of these events. During 1874, matches were held on Whit Monday and Tuesday and 20 Saturdays throughout the year. In total there were 21 matches and five handicaps. There was never any pretence about the object of these competitions, to make money, not just for the competitors and their backers but for the gambling public as well. Even events for the smallest stakes attracted betting. The amount of money at stake for all parties led to questionable actions by all concerned. Spectators were known, on occasion, to deliberately interfere with the outcome of the race. As for the competitors, they used various schemes to improve their chances. They put lead weights in their running shoes or wore clothing that slowed them down – all in order to improve their handicaps for subsequent races. It was not unknown for them to deliberately lose a race. The starting line was where most of the action took place. For example, at an event in October 1870, the 300 spectators, who had bet £300 on the race, were kept waiting for the start. It was not until "after the competitors had been a quarter of an hour at the mark – the extreme time allowed by the articles" that the referee finally ordered them to start on the report of the pistol.[61] Another race got underway after 33 false starts and another event was cancelled after 50 false starts![62] No wonder contemporary observers questioned the legitimacy of these races, although it is important to emphasize that newspaper reports frequently commented on the fairness of the races.

The increasing popularity of these commercial grounds can be gauged from the actions of the authorities. They acted, in 1875 and 1876, to close down all the commercial grounds in Northumberland because of the associated gambling. They were successful in their efforts. Dowson was found guilty of allowing betting on his grounds on June 5, 1876. By August 1876 all the commercial running grounds were closed.[63] This, of course, was not the end of pedestrianism and commercialization in the North East. In 1881, two commercial grounds, that were to become the focal points of running in the North East and East Northumberland, opened – the Victoria Grounds in Newcastle and Bebside Recreation Grounds in the coalfield. Both were opened by innkeepers and became the centres of activities in the halcyon years of the 1880s and 1890s. They became elements in a hierarchy of grounds that opened during the 1880s and 1890s. At the pinnacle, in terms of prize money and the ability to draw competitors from afar, was the Victoria Grounds in Newcastle. Within the coalfield, the Bebside Grounds became the focal point of competition and, on occasion, drew competitors from neighbouring County Durham. At a more local level impermanent grounds opened throughout the coalfield. These were multi-purpose grounds that hosted pedestrian contests on an infrequent basis.

It is the activities at the Bebside Grounds that illustrate the nature and extent of pedestrianism. In the early years the grounds were utilized by pedestrians and their backers for their contests, usually for side-stakes of £5 or £10 and usually for sprint distances: 50 yards, 110 yards, or the most popular of all, 120 yards. A hundred and twenty yards was the distance used in the handicaps sponsored by Joshua Rutter, the owner of the Bebside Grounds, after 1887. These handicaps, attracting between 32 and 144 competitors, were held at least two or three times a year. This was especially the case during the high point of pedestrianism, the 1890s. During the 1890s, 60 per cent of all challenge matches were between individuals from the same village. Another 33 per cent were from villages within two miles of each other. Pedestrianism was a local concern. One can imagine occasions at the inns where challenges would be issued after the imbibing of beer. The initial challenge was followed by meetings of the two parties and their backers, the signing of articles of agreement, the depositing of stakes, the appointment of a referee and a stakeholder, and the identification of the location of the contest. The race, itself, was simply the culmination of a complex system of negotiations and social events. The increasing popularity of pedestrianism was evidenced, in the 1890s, by the opening of other grounds that promoted athletic contests: Ashington Recreation Ground, run by colliery officials; Blyth Recreation Grounds, run by businessmen; and Bedlington Station Recreation Ground, operated by a committee of miners. All were run as commercial operations. The importance of pedestrianism is perhaps illustrated most vividly at the funeral of James Reay of Shiremoor, a leading pedestrian, who was killed in the pit at Backworth on May 25, 1895. The funeral attracted over 5,000 mourners – a vivid illustration of the importance of runners in the lives of many miners.[64]

It is apparent that pedestrianism was alive and well in the coalfield. However, none of these commercial venues provided regular organized competition. The availability of competition varied from year to year with no regular schedule of events around which athletes could plan their year's activities. The development of a regular circuit of athletic meets came from other roots, the traditional annual fairs and festivities. From the 1850s until the outbreak of the war, the summer months were filled with village celebrations at which athletics played a central role in the day's proceedings. However, athletics was only part of more extensive celebrations with parades, band concerts, cricket and football matches, stalls selling different things, and a variety of other sports. Throughout the villages of East Northumberland, Mechanics' Institutes, Workmen's Institutes, brass bands, churches, Sunday Schools, horticultural societies, Fraternal Orders of Gardeners, Foresters and Buffaloes, temperance societies, bicycle, cricket and football clubs, branches of the Miners' Union, cooperative stores, philanthropic societies, and others, held galas and sports to raise money or celebrate community. Few of these became permanent events, the majority lasted for a year or two before disappearing. Out of all these events, there emerged a small group of permanent events which, by the 1890s, had developed into a recognized circuit of athletic games. At the same time, none were solely athletic events – they retained an attachment to the communities in which they were located. The role they played in the communities, the ways in which they provided a structure to the summer, and the changing focus of some of these events to being primarily athletic events, can be seen most clearly in Cramlington.

The development of annual festivals and celebrations came not from entrepreneurs interested in making money but from existing social groups, within the villages, who promoted an annual celebration during the summer. In the early 1860s, temperance groups used athletic sports to attract people to their cause. In June 1860, a Grand Gala and Musical Fete was held at Cramlington Colliery.[65] The athletic events followed a parade through the village to a field, lent to the organizers by the colliery owners. There they partook in "Dancing on the green, quoits, leaping and football." These traditional games, such as "Sally Walker", "Kiss in the ring" and "Bob Ridley" disappeared in the 1870s. It was during the 1870s that a degree of stability was brought to the summer. The first institution to bring some permanence to the summer activities was the Court of Colonel Carr, Ancient Order of Foresters of Cramlington who sponsored an annual sports day in August from 1868 to 1893. They were joined, in the late 1870s, by the Cramlington Philanthropic and Benevolent Society with its annual sports day in late June. This continued to be held until the late 1890s. In 1879, the Cramlington Floral and Horticultural Society held its first annual show of flowers in late August.[66] This became, in the 1890s, part of the circuit of athletic games. Although programmes varied widely, all the important meets included a sprint handicap, usually a 120 yard race, and a quoit handicap. These had become institutionalized by the early 1880s – they were the prestige events which drew the largest number of competitors and offered the largest money prizes.

By the 1890s an informal, but nevertheless regular circuit of athletic games had developed in the coalfield. This informal circuit grew out of flower shows. In 1857, Seaton Burn Horticultural Society held its first annual flower show and sports. This was the first of numerous shows that sprang up in the villages. Many of them remained primarily flower shows but many also included athletic events in their offerings. For a small number of these, the athletic events gradually moved to centre stage and, by the 1890s, their primary focus had become athletics. By the early 1890s a small group of six had developed which offered the largest money prizes and were held on different weekends during August and September. Seaton Burn was followed by Ashington (1874), Seaton Delaval (1876), Cramlington (1879), Blyth (1887) and Burradon (1894). By the 1890s, the athletic scene had attained a sharper focus with a number of opportunities being provided for the athletes in the villages to compete throughout the summer. This was reflected in the development of an infrastructure to organize and run these events. In particular some "professional" handicappers and starters emerged. From 1892 until 1914, Jos Tully of Morpeth was the major handicapper of the most prestigious events.[67] However, it is important to emphasize that the majority of these remained deeply rooted community celebrations that incorporated athletics into a wider programme of community events. All included a different mix of activities which embraced the whole community: skipping, old men's races, three legged races, sack races, shooting and even waltzing competitions.

Of course, all these festivals were not the same. In fact, a hierarchy of festivals emerged based on money prizes and the area from which they drew their participants. Some remained purely local, attracting competitors from the immediate vicinity. Such was the case with the Stannington Sports of 1890. The 120 yard handicap attracted 17 competitors from within a five mile radius of the village.[68] Others such as the Cramlington Show of 1910 attracted participants from nearly all the villages in East Northumberland.[69] The catchment area for competitors for the Blyth Flower Show 110 yard handicap extended even further into 13 Durham villages.[70] At the pinnacle of athletic competition lay the Morpeth Olympic Games that in 1913 attracted 164 entries for the 120 yard handicap and 210 entries for the 110 yard handicap.[71] These competitors came from as far afield as Scotland and Lancashire. However, it was only the Morpeth Olympic Games and the commercially driven pedestrian contests in the various grounds that focused exclusively on athletics. Throughout, the flower shows remained flower shows even though the battles between the horticulturalists and the proponents of athletics over the central focus of the day were frequently bitter and prolonged. Thus, athletic competitions played different roles; for a minority they provided opportunities to compete at the highest level. However, for the majority, athletics was part of community life used to celebrate community and raise money for worthy causes. What is noticeable about all these athletic competitions, no matter what the level, is the fact that they were isolated totally and completely from the mainstream of amateur sport.

Amateur athletics, that which is written about by historians and whose athletes are glorified as the true athletic heroes, scarcely touched the athletes of East Northumberland.[72] However, amateurism was not totally dead, finding some spark amongst particular groups in particular locations: white collar workers in Ashington, Blyth and Bedlington. Between 1891 and 1914 fourteen harrier clubs were formed and all were characterized by the brevity of their existence. Only one, Blyth Harriers (1892–96), witnessed a first anniversary. Even then they did not participate in the initiatives launched in Newcastle: the Northern Counties Cross Country Association (1894), the Northumberland and Durham Paperchase League (1900) and the Northumberland and Durham Harriers League (1905).[73] The foundations of amateur athletics in the coalfield were weak indeed and demonstrate the great cultural gulf that lay between the inhabitants of the towns/villages of East Northumberland and the middle-class athletes of the larger cities.

What was the relationship of "athletics" in East Northumberland to the athletic scene in Britain? It is apparent that amateur athletics and amateur clubs did not influence the sport in the North East. Even in Newcastle, the amateur movement was relatively weak, although there were a number of events designated amateur and open only to "club" members. Of course, club membership was the defining characteristic of amateur athletics. Even in this, East Northumberland was different from other working-class areas, since athletic clubs were formed in the Midlands and the North whose members were working class. The North East was nearly totally isolated from trends and events in the rest of Britain. Where there was some similarity with other working-class enclaves in Britain was in pedestrianism and the popular village festivals. Throughout Britain, during the summer months, men, and sometimes women, participated in annual festivities. Pedestrianism was alive and well, particularly in Lancashire and Scotland.[74] Even in these cases there is little evidence of significant interaction between the different areas; the North, Scotland and Northumberland and Durham were separate regions as far as pedestrianism was concerned. Even in Northumberland and Durham pedestrianism was locally based and inward looking. It was only in the Victoria Grounds, Newcastle that there was significant interaction. Pedestrianism was very much a local concern which reinforced the inhabitants to look inside. It was only with the growth of football that significant interaction with groups outside the area took place. Athletics reinforced the isolation of the mining communities.

Fighting and boxing

It is critically important to emphasize how important fighting was in mining society.[75] It was fundamental to life itself – miners always fought, down the mine, in the street and, in particular, on Friday and Saturday nights. It was deeply engrained in the very fabric of mining life. It was one of the most visible elements of the whole concept of masculinity. Life down the mine was dangerous and brutal – only the fittest survived. A "man" was one who could stand up for himself and was always willing to test himself against anyone else. It is important to emphasize the pervasiveness and importance of fighting in the mining villages. More than anything else fighting spoke to questions of manhood. Physical prowess defined who a man was and fighting was a manifestation of manhood.

Evidence as to the popularity of fighting flowed through the nineteenth century. The Commission on Mining Districts in 1842 reported on the prevalence on fighting. In the first half of the century some colliery owners placed a prohibition on fighting within the bond. Several court cases were brought against miners for fighting in the mine. Comments were made as to the frequency of "pitman fights" in the 1860s. Without exception these took place on pay Fridays or pay Saturdays. The pay weekends became the times when grudges were worked out and money was bet. This remained the same throughout the century. Jimmy Tabarar, a miner from New Hartley, commenting on the second decade of the twentieth century, stated that there "always was a battle on Saturday afternoon. Bare fist. Strip to waist."[76] Talking about the same time, James Wilson, from Ashington, commented that there were "rough nights on pay weekends . . . Fights were frequent. Used to lay into one another, stripped to the buff, bare knuckled – until the police intervened."[77] That these were, in some sense, organized affairs is implicit in a comment made about the 1850s. "I once heard of a pitman who always gave himself an extra clean wash on pay Friday, because he wanted to look 'decent' when he put his shirt off to fight."[78] It was this semi-organized fighting that provided the foundations of the organized boxing that developed in the 1890s.

There was, however, a more formalized and organized aspect to fighting. In the early years this took the form of bare knuckle, prize fighting. Prize fighting was rooted in the traditional challenge system and involved stakes and betting. Miners were often involved in these matches. On November 5, 1849 a fight on Blyth Links between a Seghill miner and a fighter from Gateshead was stopped by police.[79] The 1890s witnessed several matches between miners from Bedlington, Backworth and Ashington. The last one involving an Ashington miner was for a stake of £30. "The fight was a most brutal one, the bare knuckles being used."[80] At the same time, the 1890s, a different variant of boxing was being promoted in the district. This involved contests fought with gloves under Marquis of Queensberry rules.

Commercial boxing involving miners first developed within the coalfield in the 1890s. This developed in the three "larger" towns/villages – Blyth, Bedlington and Ashington. At first, the contests followed the traditions of mining sport, challenge matches, meetings at inns, and signing of articles of agreement. Some of the inns became the centres of "boxing stables". Such was the case with R. Metcalf's Turks Head Inn in Bedlington. Metcalf was the chief supporter for a stable of boxers who trained at the Turks Head.[81] By 1896 entrepreneurs were offering an evening's entertainment. The first was the Circus of Varieties in Blyth. This was followed in May 1898 by a card at the miner owned and operated Bedlington Station Recreation Ground. This card included another innovation, the first Pitmen's Championship. This was the first of a variety of Pitmen's or Miner's Championships for different weight categories that were held on a fairly regular basis. This increasing popularity of boxing was evidenced in the promotion of boxing contests within the coalfield. The colliery official-dominated Ashington

Recreation Athletic Company also promoted the sport in 1902. Such was the interest in the sport that travelling showmen found it worth their while to venture into the coalfield. In 1903, the miners of various villages flocked to the boxing booth of Professor Moor's Company when it travelled through the district. Even amateur boxing raised its head in 1905, when an amateur competition for a silver cup attracted eight local fighters to Mr Lane's Pavilion, Bedlington. However, the most extensive promotion of boxing occurred in 1911 when fight cards were held on a regular basis at Ashington Miners' Hall and the Pavilion Theatre, Hirst. Nearly all the cards at Ashington/Hirst featured Peter Brack, the highly popular local champion and, on one occasion in 1911, there was an exhibition by Jack Johnson, the world champion.

However, the real development of boxing, which provided an alternative to the mines for a small number of miners, occurred in Newcastle. In fact, the commercialization of boxing started in Newcastle in the early 1890s and involved the most important figure in the promotion of boxing – Jimmy Lowes. In 1891, Lowes opened the Percy House Boxing Saloon in Newcastle. Conditions were not propitious for commercial boxing and it succumbed after a brief existence. In 1900 he opened two other saloons. This was the beginning of regular cards. This led to a demand for boxers. Many of these came from the mining villages. From 1901 the various Pitmen's or Miners' Championships moved to Lowes' emporiums mainly because he offered the largest purses. Relatively regular cards were offered throughout the early 1900s until Lowes closed Ginnett's Circus in 1908. In fact, this was simply the prelude to the great era of boxing in the North. In September 1909, St James' Hall, the first custom built boxing hall with a seating capacity of 3,000, opened for business. From thenceforth, boxing cards were offered on a weekly basis on Saturdays and Mondays – boxing had been institutionalized. For the next 60 years St James' Hall was the mecca of boxing enthusiasts throughout the North East of England.

The involvement of pitmen at St James' Hall provides insights into both the commercialization of boxing and the opportunities for a limited number of pitmen to escape the hardships of the mine. The first to fight in a championship at Newcastle was Joe Mather, who fought J. Keeble of Choppington over 20 rounds for the nine stone Pitmen's Championship at Ginnett's Circus in June 1901.[82] Like the majority of pitmen fighters Mather disappeared from view after 1901. More informative is the career of Tom Moore, a pitman from Seghill, who fought between 1906 and 1912. Between 1906 and 1908, Moore fought 11 times at the circus, an average of four times a year. This reflected the state of boxing, ever present but lacking the weekly regularity that came with the opening of St James' Hall in 1909. In the 21 months from the opening of the hall until June 16, 1911, when he sustained an injury, Moore averaged more than one bout a month.[83] These 24 bouts were all 15 or 20 round contests. In order to meet the demand for boxers, as a result of the weekly contests at St James' Hall, fighters were forced to fight on a monthly basis. Moore was not atypical. Peter Brack, the Ashington miner, participated in 20 contests in 1911 and 1912. Jim Berry, of Dinnington, one of the best fighters to come out of East Northumberland, fought 146 rounds in nine bouts between February 19 and November 22, 1912. Moore, Brack, Berry, and many others provided a sizeable portion of the supporting bouts for the main event at St James Hall. In fact, in 1912, 19 miners graced the ring at the Hall. For this small group, the fight game provided opportunities to get away from the mine, albeit exchanging one form of exploitation for another.

How does boxing in East Northumberland relate to the history of boxing in Britain? In very general terms, it developed at the same time in the 1890s and early twentieth century. Additionally it was very much a working-class sport. The pitmen of the villages joined workers from the working-class areas of the cities of Britain as avid supporters of boxing.[84] However, there were greater differences than similarities. The infrastructure of boxing was significantly different. Amateur boxing based on clubs associated with pubs, churches and working men's clubs that defined the sport in London and other cities was not present in the villages.[85] There is no evidence that the London based Amateur Boxing Association and the regional affiliates in the Midlands and the North had any presence in the North East. Boxing in East Northumberland remained tied to the traditional challenge matches until they petered out early in the 1900s. It was only the professional variant of boxing that prospered in the mining villages. Even then it lagged behind its development in London and other cities. The regular weekly shows that developed in London, Birmingham and Manchester in the period 1887 to 1897 did not develop in Newcastle until later.[86] While cards were presented in the 1890s, it was not until 1909 and the opening of St James' Hall that cards were offered twice a week on a regular basis. Like so many sports the history of boxing in the villages of East Northumberland did not fit in, except in general terms, with what went on in the rest of the country.

Conclusion

This chapter has dealt with the relationship between the old and the new, the ways in which tradition moulded and influenced sport in the late nineteenth century. Additionally, it has focused on the changes that undoubtedly took place; the games of 1914 were different from those of 1850. What did the new sports owe to tradition and what did they owe to the changing society? Traditional games and practices remained stubbornly in place. However, they were practised by an ever diminishing coterie of players whose formative experiences were in the mid-nineteenth century. At the same time traditional sports remained alive and well – cricket, athletics and pigeon flying experienced significant growth. Yet cricketers still preferred exhibition and friendly games to league competition. Climbing the greasy pole still appeared on some programmes. Short distance pigeon flying retained its hold on the majority of fanciers within the colliery villages. What, then, changed? The major change lay in the form of competition. Traditionally miners competed in the ubiquitous challenge matches within the infrastructure of social interaction at the inns and the predominant role of the competitor and his backers in arranging the event. By the turn of the century these had become the exception rather than the rule. Handicaps involving large numbers of competitors, pre-scheduled league competitions, and mass contests came to typify sport. Basically there was a great increase in the number of competitions and competitors. These changes precipitated two inter-related but separate processes that changed the basic meaning of sport – the processes of organization and commercialization.

During the second half of the century clubs came to characterize sport. Short-lived rowing clubs emerged in the early 1860s. The first homing clubs were formed in 1877. With the formation of clubs, decision making was often placed in the hands of a small group of people; gone were the days when competitors and their coterie of backers, supporters and hangers-on met at an inn to thrash out the terms of the contest. Now, decisions were made by club officials. However, this process of centralizing the decision-making process was not an unqualified success. Challenge matches continued to be made, short distance flying remained steadfastly independent. Only the long distance clubs, run by men with some administrative experience, moved into district and national organizations. If miners wished to compete at the highest level of competition they had to affiliate with groups whose decision-making power lay outside the mining districts. The isolation so conducive to solidarity was gradually being eroded. Of course, the growth of regional and national organizations was predicated on the stability and permanence of clubs. This did not characterize mining sport.

In fact, sport in the villages was characterized by impermanence, the majority of clubs having a brief existence before disappearing. Sport paralleled life, being characterized by uncertainty, instability, and periods of feast and famine. It was only those sports that drew from people with some organizational experience that acquired a degree of permanency. This brought sport within the purview of the middle class and those miners who had experience of organizing. Additionally, until the war, the highlights of the year focused on the annual celebrations rather than regularly scheduled games. However, towards the end of the century a degree of regularity was creeping into sport. Sport was moving into the world of modern sport.

The other process that influenced sport was that of commercialization – the selling of sport for a profit. Since mining sport was always related to money it is difficult to disentangle the complex processes. Commercialization, in its fullest sense, was only seen in boxing. Even at this early date the ugly side of commercialization, that of exploitation, reared its ugly head. The excessive demands made on boxers attest to the exploitative nature of commercial boxing. At the same time, boxing did allow miners to follow an alternative occupation. It provided chances for a new life. Thus, full fledged commercialization was a two-sided coin. Of course, commercialization was not only concerned with making money. For many innkeepers sport was the *raison d'être* of their lives. For example, Robert Metcalf of the Turks Head Inn, Bedlington promoted boxing by sponsoring a stable of boxers, providing the training facilities and providing the financial backing. He also promoted the prestigious Bedlington Athletic Sports during the 1890s. His name also appeared on the subscription lists for cricket clubs and his involvement in pigeon flying was as an active participant in the sport. Thus, we must be careful in assessing the degree and impact of commercialization. Yet one of the consequences of the growth of organizations and the increasing size of purses was to make financial viability increasingly important. While the objectives of the activities may not have been to make money, increasingly a failure to cover costs could spell disaster. For cricketers, money was rarely an issue as they were given financial support by the land and colliery owners. The other sports were not so lucky and were increasingly tied to financial viability. With increased organization and extended competition, costs rose – it cost to send pigeons to Rennes, France, to build specialized rowing skiffs, and to frame handicaps for athletic events. Thus, in one sense, sport had to become commercialized in order to survive. Money became the rock upon which many clubs floundered – thus, the endemic instability of mining clubs and sport.

Perhaps the more important question relates to how this history fits into what was going on in the rest of Britain. While the processes of commercialization and organization were common and even though all sports were to be found throughout the country, their histories were very different. In cricket, the history in East Northumberland found resonance in every part of the country – elements from the North and South, country house cricket and league cricket, and teams sponsored by particular groups. With respect to rowing, abortive attempts were made on the Wansbeck and Blyth to commercialize the sport and create amateur clubs. Amateur athletics which took root in working-class areas across the country found no foothold in the East Northumbrian villages. Even in boxing, the history was different from that experienced in London and other cities; there were no boxing clubs in the mining villages. East Northumberland was very different from anywhere else. The same games had different histories – cricket in Lancashire was not the same as cricket in East Northumberland, the history of boxing was significantly different than elsewhere, rowing was inward looking and athletics eschewed amateurism. Thus, the same games exhibited different forms and were used for different ends.

5 New sports for new times

This chapter deals with a group of eight sports that were, on the surface, notable more by their differences than their similarities. They were held together, loosely, by four attributes. In the first place, unlike the sports in the previous chapters, tradition played a small role in their development. While some had histories in other parts of the country, all were new to East Northumberland. Additionally, seven of the eight sports developed within the context of formally constituted clubs, a radical departure from the traditional methods of organizing sport within the mining villages. The emergence of club-based sport is one of the defining characteristics of modern organized sport. This development of clubs was accompanied by a shift from being exclusively a sporting competition to increasingly including and emphasizing the social aspects. Finally, and providing a common thread, is the importance of the early 1890s in the histories of all these sports. Six of the eight sports could never be regarded as anything but minor sports which appealed to very small segments of the population of the towns and villages of East Northumberland. Two of the sports, bicycling and billiards, were significantly different from the others in that they attracted large numbers of participants and spectators. At the same time, even more than the other sports, the period 1890 to 1895 was critical to their histories.

Swimming and fishing

Small groups of aficionados formed swimming and angling clubs in the 1880s. Although swimming competitions were held in conjunction with some of the regattas in the 1860s, the competitive swimming clubs were formed in the coastal villages of Seaton Sluice (1890) and Newbiggin (1893).[1] Perhaps more important were the clubs formed for health reasons. By 1895 there were seven clubs whose members regularly took morning "dips" for health reasons. These clubs were strongly supported by colliery officials, ministers and doctors. However, miners were involved in four of the clubs and one of the acknowledged leaders was John Bryson, President of the Northumberland Miners' Union. Four of the clubs were specifically designated "amateur".[2] The swimming clubs were joined by five angling clubs, the first being the Choppington and Guide Post Conservative Working Men's Fishing Club (1880). The five clubs formed in the 1880s were devoted to fly fishing

Plate 5.1 Stakeford Swimming and Rowing Club, *c.*1912.
Reprinted with permission from Martin, E. (1997) *Images of England: Bedlingtonshire*, Stroud: Tempus Publishing

on the local rivers.[3] The two that developed in the 1890s were different in that their membership was over 100 and they fished the border rivers. Who these fishermen were is difficult to determine but the evidence suggests that miners were involved.

Golf, tennis and lawn bowling

Golf, lawn tennis and lawn bowling were different from any of the previously mentioned leisure activities. From the very beginning they focused as much on the social elements as the competitive – the club house and the garden became the focal point of an extensive social life. During the winter the clubs promoted social activities with teas, suppers and dances. While their members were drawn from the upper echelons of East Northumbrian society there were differences in the three sports that illustrated the way in which appearances were sometimes deceptive.

Golf made its appearance in the early 1880s on the expanse of Newbiggin Moor. It was local landowners, ministers, doctors and businessmen who formed the Newbiggin Golf Club (1880) that played on the links of Newbiggin Moor. It was unashamedly elitist. However, in an unusual move, the members of the Newbiggin Golf Club set about forming another club in Newbiggin. As a result of their efforts the Newbiggin Workmen's Golf Club was formed in 1884 and held its first competition on Newbiggin Moor in March 1885.[4] There is no evidence of inter-club competitions and certainly not between the two clubs – the social divisions were rigidly maintained. In 1905 some Blyth businessmen met to form Blyth Golf Club.

Like Newbiggin G.C. the Blyth club played on a links course on Blyth Links. By 1912, there were five golf clubs in the area. The creation of the socially acceptable Blyth club lead to the beginnings of inter-club competition between Blyth and Newbiggin. This ushered in a new era in the history of golf in the district – the development of extensive inter-club competitions. In fact, the Newbiggin G.C. was one of the earliest in England and its promotion of a working man's club made them pioneers.

Lawn tennis, too, owed its origins to the upper echelons of Northumbrian society although its roots soon spread much further afield than did those of golf. Even more than golf, the social elements of lawn tennis were at the heart of the club's *raison d'être*. This was reflected, most graphically, in the inclusion of women. While their social role was important, women also graced the courts and eventually the administrative bureaucracy. Many of the "promoters" of lawn tennis were "foreigners" to the area: men who came to live in East Northumberland because their occupations brought them there – doctors, clergymen, colliery officials and teachers. Again the first to organize were the businessmen of Morpeth, Blyth and Seaton Delaval in the late 1880s. In fact, Seaton Delaval Lawn Tennis Club held its second annual ball in January 1889 in the Colliery Schoolroom – 60 couples attended and it claimed to be the only tennis club in the district.[5] We obtain a clearer view of the roots of the game in the creation and development of clubs. Ashington Lawn Tennis Club was formed, in 1893, by a group of young men attached to the Young Men's Mutual Improvement Society with strong support from the highest officials in the coal company.[6] Anglican vicars played a central role in the formation of clubs in Bedlington, Newbiggin, Choppington, West Sleekburn, Hirst and Bothal. Frequently acting as presidents of the clubs, playing on the teams, and offering the vicarage grounds for tennis, the vicars played a pivotal role in the growth of the game. The Anglican vicars were given support by the local landowners who frequently provided grounds for the clubs. Such was the case with the Bedlington Lawn Tennis Club who played on grounds provided by the local "squire", E.A. Burdon of Hartford House. Similarly, Lord Hastings gave permission for a club to use the grounds at Seaton Delaval Hall. The relationship between tennis and the social elite was clear. Until the late 1890s, lawn tennis was limited to Blyth, Bedlington and Ashington. The focus was predominantly on intra-club competitions with the social activities being just as important as the games. Things changed in the early 1900s. There was an increase in the number of competitions and clubs. Between 1907 and 1914 nineteen new clubs emerged in the district. The members were drawn from the professionals and businessmen of the communities. Hardly a weekend passed without a competition somewhere in East Northumberland. In the majority of cases these matches involved both men and women – the social side was never forgotten. Increased interest in competition culminated in a meeting, convened on September 15, 1911, to consider the formation of a league. The East Northumberland Lawn Tennis League, formed at that meeting, comprised seven teams. It commenced play in May 1912.[7]

Somewhat surprisingly, although miners were, as far as can be ascertained, absent from these clubs, there is evidence of miners being involved in the game from a relatively early time. The first unequivocal evidence of their involvement was the formation of the short-lived club at Newbiggin Mechanics' Institute in 1895, by the Rev. J. Nichol Grieve MA.[8] In April 1896, the management committee of the miner-owned Bedlington Station Recreation Ground, on the advice of Mr Hare, the colliery manager, voted to lay down tennis courts.[9] In 1900, the committee of East Holywell Workmen's Institute introduced the game to their members. Within three months the new club boasted 50 members.[10] The clubs within the mining villages were, in some respects, different; for example in 1907 the president of Shiremoor Lawn Tennis Club was Miss Hardy and the secretary Mrs Veitch, a teacher at Shiremoor Council School. The influence of those outside the mining communities was illustrated, most clearly, in the introduction of the game to Shiremoor public schools in 1910, perhaps through the efforts of Mrs Veitch. For a brief period of time, the local schools played "friendly inter-school competition."[11] What is clear is that the involvement of the miners was never on a long term basis. Rather, it appeared to be a fad which emerged and then disappeared. At the same time, their involvement does serve as a warning against a simplistic view of what miners did or did not do with their leisure time.

Participation in both golf and lawn tennis was limited by access to facilities. In the case of golf, access was gained to public land while, for lawn tennis players, courts were provided by landowners, vicars and colliery officials. Lawn bowling, too, required specialized facilities. In this case the first greens were provided through public funds and it was the opening of Ridley Park, Blyth in 1904 that ushered lawn bowling into East Northumberland. A year after the opening of the park a group of Blyth businessmen formed the Ridley Park Lawn Bowling Club.[12] For the first seven years they participated in intra-club activities and it was not until the formation of Blyth Excelsior in 1912 and Blyth Masonic in 1913 that they moved into inter-club competition. In the same year, a new green costing 200 pounds was opened near Bedlington Station by Dr Tindall, the president of the new Sleekburn Bowling Club. These represent the tentative beginnings of the game that was to spread like wildfire through Northumberland and Durham during the inter-war years.

Cycling

At 3.30 on the morning of June 10, 1876, James Bowman, the sub-postmaster at Cramlington left for Edinburgh on his newly acquired 50" ordinary, better known as a penny farthing bicycle.[13] Eighteen hours later he arrived at his destination after traversing 119 miles of pot-holed roads and tracks that linked Cramlington and Edinburgh. Little did Bowman realize that he was ushering in a new era that would eventually revolutionize the working man's physical horizons. From the outset, the bicycle was used for sporting and leisure purposes. In fact, bicycling got its start, in East Northumberland, as a competitive sport. This reached a highpoint in the early 1890s before entering a period of precipitous decline, eventually becoming a minor sport. In terms of more broadly based leisure activities, bicycling "took off" with the invention of the safety bicycle and the pneumatic tire. This occurred in the late 1880s and it too reached its apogee during the 1890s, although it maintained a more robust presence until the war.

Fifty-inch Ordinaries were introduced into the colliery districts in 1873. By 1876, there were enough "high wheelers" around to enable a small group of enthusiasts, in the Cramlington area, to form Cramlington Amateur Bicycle Club, the second in Northumberland and one of the most successful clubs in the pre-1914 period. During the next 10 years another eight clubs were formed but only two lasted for more than two years. At the same time, there was increased interest in competitive bicycling. This was evidenced by the increased number of competitions during the summers, and the involvement of bicycle manufacturers, innkeepers and colliery officials in promoting the sport. The most popular form of racing until 1891 was road racing. On September 8, 1877, John Battensby of Hartley, a future world champion, and Jeremiah Cleminson of Cramlington, a 21-year-old miner, both of whom became leading professionals in the 1880s, raced over four miles from the Victoria Inn, Seaton Terrace.[14] The interest of innkeepers and bicycle manufacturers precipitated the question of professionalism to the forefront. This was to remain a bone of contention until its solution in 1895. However, in these early years, professionals and amateurs competed on the same afternoon, on the same track, without fuss or fanfare, and for the same prizes – money.

While road racing remained popular, it was not ideal for spectators and the popularity of the sport can be gauged by the attempts to promote track racing on enclosed tracks which allowed for entrance money to be charged and also allowed for a better spectacle. By the early 1880s several innkeepers constructed tracks for competitive bicycling in Seghill, Blyth and Bebside. There were, however, a number of grounds that did achieve a degree of stability and success. These grounds became the centres of racing within East Northumberland. The initiative for the construction of these tracks came from the bicycle clubs themselves. This, then, leads us once again to the symbiotic relationship between successful initiatives and approval of the colliery officials. On September 11, 1880, Bedlington Amateur Bicycling Club inaugurated their new track. "The owners of Bedlington Colliery, with their usual liberality have provided an excellent bicycle track of 8 laps to the mile, in the centre of the village for the use of the workmen employed at the colliery."[15] This was followed by the opening of tracks in Bedlington, Burradon, Cramlington and New Hartley. It was these tracks and these villages that remained at the heart of cycling in the coalfield.

As the popularity of the Ordinary declined and safety bicycles became available to a wider segment of the population the number of cyclists on the roads increased. The increase in bicycles shifted the focus of cycling from racing to activities that involved larger numbers of men and, on occasion, women. Two events in 1887 signalled the beginning of a new era in cycling. In May 1887 the Blaydon Amateur Cycling Club promoted the first church parade to the grounds of Gibside Hall.[16] This was the beginning of a number of church parades that were held throughout the summer months. On a Saturday afternoon in July, seven members of the New Hartley Bicycle Club rode over "shocking" roads to Sheepwash.[17] This was the first of the Saturday and Sunday runs that became very popular during the 1890s.

At the same time, the period 1888 to 1895 saw bicycle racing reach its greatest heights of popularity before undergoing a rapid decline to become a minor sport.

By any criteria, the early 1890s were the halcyon years of cycling. The growth in popularity was directly related to the development of the safety bicycle with pneumatic tires and the mass production of bicycles. Suddenly, bicycles fell within the purview of a much larger segment of the population. This resulted in an increase in the number of riders, both for pleasure and racers, on the inadequate roads of East Northumberland. This sudden increase in traffic on the roads brought complaints about "scorchers", individual riders going so fast that they threatened the safety of pedestrians, other riders, and other road users. This increased activity aroused public concern over the "scorchers." Consequently in 1891, the County Council passed a motion that banned racing on the highways of Northumberland.[18] This resulted in cases against racers being brought to trial at the Moot Hall Police Court, Newcastle and Blyth Court.[19] The result of these actions was to change the focus of racing. In 1890 over 50 per cent of challenge matches were held on the county roads. By 1891 racing had disappeared from the roads and moved onto enclosed tracks. The shift to track racing also shifted the focus from individual challenge matches to club-sponsored handicaps and club events. Thus, the actual nature of cycling changed. This was assisted by ownership of the tracks. All the successful tracks of the 1890s were built and run by cycling clubs. Thus, the world of cycling was now played out within the context of community developed clubs. It was the nature of these clubs and their fragile financial foundations that provided the conditions for their failure in 1894–95.

Forty-one clubs were formed between 1888 and 1895.[20] The majority of these lasted for less than two years. However, there was a core of 17 that provided some stability and focus to the world of cycling. The majority of these clubs focused on seven tracks, five of which lay within a two mile radius of Seaton Delaval. Of the 41 clubs, 21 lay within this area, eight of the permanent clubs and five of the tracks which were the focal points of competitive racing. The two other centres were Ashington (1886) and Bedlington (1893). It is in the histories of these tracks that we get some sense of the reasons for the rise and fall of competitive racing. The nature of these clubs and the reasons for their demise can best be illustrated through the history of one club, the New Delaval Black Diamond Amateur Bicycle Club, formed in 1886 and dissolved on February 23, 1895. Its history spanned the real beginnings of the sport of cycling with the advent of the safety bicycle to its height of popularity in the early 1890s and its precipitous decline by 1895.

The history of the New Delaval Black Diamond Amateur Bicycle Club illustrates all the basic ingredients underlying the sport of cycling in the mining villages.[21] It saw the light of day at a meeting held at Mr Lacey's Black Diamond public house in 1886. The relationship with the inn and innkeeper was to remain, despite the renaming of the club in 1888 as the New Delaval Amateur Bicycle Club. This renaming revealed an important aspect of bicycling in the villages; they were very concerned about amateur status. While the outside world rarely considered them amateurs, they certainly regarded themselves to be amateurs. The Black Diamond remained the centre of the extensive social life of the club. During the winter, regular socials and concerts were held. Additionally the annual meetings, monthly meetings of the executive, and special meetings of the members were held at the inn. Also, the club visited other clubs and inns in the surrounding area, giving concerts. However, its *raison d'être* was cycling and the promotion of the sport. Although segments of the membership did get involved in the weekly runs and the annual church parades, the primary focus, throughout, was racing. And it was the desire to promote racing that established the first priority of the club when it was formed in 1886: the acquisition of ground for an enclosed track. After nearly a year of negotiations with the coal company, they were granted the use of a field and provided with some financial assistance. R.E. Ornsby Esq. of Seaton Delaval House, the highest on-site official of the company, agreed to be president of the club. Thus, one of the basic necessities for success had been established, a close relationship with the coal company. The close relationship between the club and the coal company remained throughout the club's existence. Through the physical labour of the members and the raising of £60 to pay for the work, the enclosed track was eventually opened in 1889. One of the first events was the visit of five American lady cyclists, who competed against the local champions with a fair degree of success. Miss Lottie Stanley, an 18-year-old professional, from Pittsburgh, Pennsylvania, despite giving the local male champion, G.H. Symington, a miner, a 28 yard start, still managed to beat him by a foot.[22] With the gradual shift to track racing the New Delaval track hosted numerous club-sponsored events. Between April and September, the club members participated in weekly two mile, one mile and half mile races and handicaps. Additionally, during the period 1890 to 1893, they hosted, annually, four or five meets which attracted racers from all over the North East. This increased popularity was a mixed blessing since the members of New Delaval travelled to other meets for competition. They tried to generate income by renting the track out to other cycling clubs but this met with limited success. The 1893 season was a disaster as revenue from racing declined. The executive of the club attempted to meet the situation by promoting pedestrian matches in October. This initiative met with some success but the bottom had fallen out of racing. In 1894 there were very few races in the district and on February 23, 1895 the club disbanded and, appropriately, turned the ground over to Newsham Villa Football Club.

The history of the New Delaval club was typical of racing clubs in the district; relatively regular competitions, visits to other clubs, but insufficient use of their own ground to provide an adequate financial return. This meant that when support for the meets declined the clubs were precipitated into a financial crisis that, in many cases, led to dissolution of the club. However, while the success of racing within the district was dependent upon support from the colliery owners and officials, the reasons for the decline and near extinction of racing lay outside of and beyond the power of the miners and their supporters. We have already seen how the actions of the County Council to restrict road racing virtually eliminated road racing by the end of 1891. The decline of track racing was somewhat different. This was related to the wider world of bicycle racing and the problems facing the National Cycling Union (NCU) in their attempts to bring some order to the world of cycling.

At the heart of their problems was the vexatious question of the relationship between amateurs and professionals. It was unfortunate for East Northumberland riders that the campaign to adopt a most rigid and exclusionary definition of an amateur was led by the Newcastle Branch of the NCU, the branch with which East Northumberland riders had to affiliate if they wished to compete in the prestigious amateur meets, the entry point into the highest level of amateur sport. This demonstrated, vividly, the problems facing East Northumberland clubs, in any sport, when they ventured into competitions run by externally controlled organizations.

The problems of the 1890s were inextricably linked to the explosion of interest in cycling that commenced in 1891 and peaked in 1894. In efforts to solve the problems of unwanted riders at the "amateur" meets the NCU, led by the Newcastle Branch, proposed two related courses of action, the enactment of a rigid definition of an amateur and the imposition of a licensing scheme for all amateur riders.[23] Within the coalfield clubs the question of differentiating between amateur and professional, although important, was never used as a system of class discrimination as it was in the Amateur Athletic Union and later the NCU. Several clubs in the district, consisting of miners, designated themselves as "amateur" clubs. At the same time professionals and amateurs competed on the same tracks in the same meets. Additionally riders frequently shifted status from amateur to professional and back. Like the amateurs of the NCU, it was prize money or payment for riding that determined amateur or professional status. Unlike the NCU, within the coalfield this only applied at particular moments. The barriers between amateur and professional were porous, allowing individuals to shift back and forth. This liberal approach to the question of amateur status did not cause any problems as long as the competitions remained within the district. It was when riders desired to compete in outside meets and local clubs attempted to attract the top riders from outside that the problem became severe. It was during the period 1891 to 1894 that the battle lines were drawn and the battle fought. The victory of the NCU in 1894 was, in some senses, a pyrrhic victory – they won the war but in so doing condemned bicycle racing to the status of a minor sport.

The NCU was formed in 1882.[24] The Newcastle Branch, dominated by middle class Novocastrians, remained relatively small throughout the 1880s. By 1889 there were 21 clubs affiliated with the Newcastle Branch of the NCU, none of which were located within the mining villages. It was not until 1890 that Burradon Cycling Club affiliated with the NCU. They were joined, in 1892, by clubs from Choppington, Blyth and Bedlington. This paralleled the beginning of a brief period when it appeared that cycling would become a major sport. Membership in the Newcastle Branch of the NCU skyrocketed to an all time high of 109 in 1894. As the popularity of competitive cycling soared so did the problems facing the NCU, in particular, the thorny question of the relationship between amateurs and professionals. This was precipitated, in part, by the affiliation of clubs from the mining villages whose approach to the question had always been liberal. It was not surprising, therefore, that soon expressions of concern were voiced. It was in 1892 that the first expressions of dissatisfaction with the Newcastle Branch were voiced over the "Makers

Amateur" question and the perception of Newcastle authoritarianism. This did not bode well for the future. The makers amateur question dominated the NCU annual meeting in 1892. The actions of the Newcastle Branch in 1892 precipitated a "Cyclist Revolution" in the colliery districts, The Barrington Amateur Bicycle Club called a meeting at the Railway Hotel, Choppington to consider "The arbitrary manner of the NCU, and to take into consideration the licensing of racing cyclists."[25] The meeting revealed that the problems were rooted in different ideological views about amateurism and the strong resentment of the way in which Newcastle attempted to get its own way. The discussion at the meeting indicated that the institution of exorbitant handicapping fees, the new licensing scheme and the cost of framing handicaps would mean the death of racing in the district. The colliery clubs, comprising miners, just could not afford the increased costs. They attempted to run meets independent of the NCU; the meets failed. During the summer of 1893 the annual club meets were failures. For example, the meet of NDABC, in 1891, attracted over 150 riders for the two events.[26] Two years later the same meet attracted only 40 riders. The bottom had fallen out of racing. This was the beginning of the end. On December 9, 1893, the Bates Cottages Amateur Bicycle Club collapsed due to the failure of its meets.[27] By the end of 1895 another seven racing clubs had collapsed due to financial failure.[28] Although the innkeepers of Blyth made an attempt to keep racing alive, they failed. The flame of NCU-controlled racing was not totally extinguished, it remained flickering in different parts of the coalfield with five or six clubs maintaining membership in the NCU. The major "stronghold" was Ashington where the "amateur" Portland Cup competition was held on an annual basis. This competition attracted racers from Newcastle, Durham and a few colliery-based clubs in Ashington, Bedlington and Barrington. The NCU applied the final *coup de grâce*, in 1896, when they voted 39 to 26 not to allow amateurs and professionals to compete at the same meets. Interestingly, of the 36 clubs formed after 1895, not one included amateur in the title of the club. Bicycle racing continued to maintain a small presence in the district but by 1913 there were only 13 licensed riders in the district and only 49 affiliated clubs. Bicycle racing had become a minor sport.

While racing was the focal point of cycling for a small but enthusiastic group of riders, the majority of club members did not participate in racing. There was another, and in terms of number of participants, more important dimension to club life: the weekly club rides and the less frequent cyclists' parades. On July 23, 1887, seven members of the New Hartley Bicycle Club met at their headquarters before embarking on a Saturday afternoon run to Sheepwash. They successfully accomplished their task. At first, the club rides emanated from the competitive cycling clubs and provided another dimension to club life, but in 1890 the Holywell Touring Club, a club whose whole focus was upon club rides and club activities, was formed.[29] From that date the majority of new clubs were touring clubs. At the same time the basic format of club activities had been institutionalized. The season started on Good Friday and continued every weekend until the end of September. By the early 1890s every weekend witnessed squads of riders from different clubs riding along the roads of Northumberland to such destinations as the local beauty

Plate 5.2 Seaton Sluice Cycling Club, *c.*1905.
Reprinted with permission from Allan, T. (1992) *Seaton Sluice and Old Hartley: A History in Photographs*, Morpeth: Northumberland County Library

spots of Sheepwash, Stannington and Newbiggin. Invariably they stopped for lunch at a local inn where they partook of refreshments. The number of riders varied from 20 to over 100 and the distances from 10 to 60 miles in a day. These rides were not just an unorganized crowd of riders charging along the roads of the county. Of course, like the "scorchers" their presence on the roads frequently attracted the unwelcome attention of the police and authorities. The clubs reacted by taking actions to control the "hordes" of riders. They became highly organized events with recognized leaders, buglers, and "whippers in". Groups of 100 riders, all decked out in club colours, streaming through the colliery villages must have provided a vivid spectacle. While club riding also reached its apogee in the early 1890s, it maintained a strong presence well into the twentieth century. In fact, 36 new touring clubs were formed between 1896 and 1914. However, these were formed in limited locations: 20 towns/villages. Interest began to decline after 1905 and organized cycling became the preserve of a small band of enthusiasts.

There was another, less frequent, activity that entailed club members joining with other clubs to parade through the towns and villages. The majority of parades were church parades. On May 24, 1891, 600 riders, representing 17 district clubs, with 1,400 non-riders, attended a church parade at Seaton Delaval Hall.[30] The unexpected success of the parade led the organizers, Holywell BC, to promote another parade in the following year. Forty-nine clubs assembled in Seaton Delaval and, lead by the Seaton Delaval Brass Band, paraded the two miles down the avenue to the hall. The parade, which attracted great attention in the village, made £36 for charity. However, the day was marred by some objectionable behaviour on the part of some of the participants. The bishop's speech was interrupted by unnecessary epithets and an "idiotic buglar" [*sic*] who interrupted the proceedings.[31] As a result Lord Hastings withdrew his permission for the grounds to be used for the parade and thus it disappeared. The following year witnessed the inauguration of the only church parade to survive until the First World War – the Stannington Parade. On May 5, 1893 18 clubs from the colliery villages rode to Cramlington where they joined together to parade the three miles to Blagdon Hall, the home of Sir Matthew White Ridley.[32] At Blagdon Hall, they were joined by 37 other clubs who had gathered in Newcastle and paraded up the North Road to Blagdon Hall. In the grounds at the Hall, the cyclists were joined by 5,000 spectators who had travelled by other means to take part in the service which was held in the grounds of the Hall. The stream of cyclists converging upon Blagdon Hall must have provided a never-to-be-forgotten sight. In the twentieth century the parade still attracted a significant number of riders. In 1910, 3,000 cyclists from 60 clubs converged upon the grounds at Stannington. Illustrating its roots in the mining communities, the parade was postponed in 1912 due to the Miners' Strike. It was held at the end of July when the strike was over.

While Stannington was the only parade with any degree of permanence, there were other parades that flourished for a brief time before disappearing. The most important of these were four parades which were held on the grounds of local landowners. The parade to Seaton Delaval Hall has already been mentioned. From 1896 to 1899 the Bedlington clubs held a parade to Hartford Hall, the home of the Burdons, the "squires" of Bedlington. In 1899 and 1900, Ashington and District Cycling Club, led by Ashington Orchestral Band, paraded to Bothal Castle, the Duke of Portland's estate. From 1901 to 1903, the newly formed North Seaton Cycling Club paraded to North Seaton Hall. All these events were gala days in the summer calendars of the socially prominent. In each case they attracted substantial numbers of riders and spectators. However, except for Stannington, interest soon waned and they disappeared.

The parades to the grounds of the local landowners were the highpoint of the cycling parades. However, there were other more locally focused parades held to raise funds for churches or some charity. Parades were held at Blyth (1895), Seaton Delaval (1898), North Seaton (1898), and Seaton Burn (1900). The parade organized by Hartley Lily Temperance Cycling Club on August 20, 1898 typified these parades. Two hundred cyclists associated with three temperance cycling clubs met at the Moor Edge at the north end of Seaton Delaval.[33] Led by the Dudley UMFC Brass Band they paraded through the village to a field granted by a farmer. There they sang hymns and made a collection for the Boys' Fresh Air Fund, the Girls' Home, and the widow of a Mickley miner killed while returning from the Stannington Parade. On occasion clubs from the district travelled considerable distances to attend parades held in aid of some charity. For example, in 1903, 14 district clubs rode to Alnwick to participate in a church parade in aid of Alnwick Infirmary.[34] This was the last recorded church parade.

Who were these cyclists who, for a brief period of time, raced and rode on the roads and tracks of East Northumberland? During the era of the penny farthing,

1876 to approximately 1890, for most people, the cost of the vehicle was prohibitive. Those early riders that can be identified were, for the most part, not miners. The Bowman brothers of Cramlington were postmasters, W.H. Endean was the parish clerk of Cramlington, and J.E. Jobling, one of the most successful of the early riders, later became the agent of the Bedlington Coal Company. At the same time, there is evidence that from as early as the 1880s miners did participate on the racing track. John Battensby and Jeremiah Cleminson, the most successful of all riders, were miners. The track developed with the permission of the Bedlington Coal Company was specifically created for the "Use of workmen employed at the colliery."[35]

The development of the safety bicycle in the late 1880s ushered in a new era, an era in which the bicycle became available to a wider segment of the population. Of course, some riders held on to the past. As late as the early 1900s, Jack Dorgan was impressed by Mr Battensby; could it be the same John Battensby? Dorgan commented on him riding "A penny farthing cycle and I shall always remember him in his cycling clothes, a Norfolk style jacket, knicker bockers fastened with a buckle at the knee" – surely an unforgettable sight![36] Many of the clubs that developed in the early 1890s were organized and run by miners. Such was the case with the Astley Club of Seaton Delaval that was formed on June 20, 1890 at the Astley Arms Inn, Seaton Delaval. The development of this club was rooted in the traditional relationships with the inns and Mr Dawson, proprietor of the inn, played an active role for many years. The club itself was the brainchild of two young miners, Joseph Beaney, a 20-year-old, single miner and Matthew Stoker, a 23-year-old, married miner. The club was, for the most part, composed of young, single coal miners. Of 15 that could be identified, 12 were miners and the others were a railway signalman, a butcher's apprentice, and a blacksmith. Twelve were single and the average age was 22 with the youngest being 20 and the oldest 39.[37] That this was not an isolated event was demonstrated in 1891 at the First Seaton Delaval Church Parade. An observer commented that "By far the greater number of clubmen present were miners located in the Blyth and Morpeth districts."[38] There were 348 clubmen present representing 16 district clubs. Thus, it is clear that, in fact, the majority of clubs in the district involved the miners as active participants both on the tracks and in an organizational capacity. Certainly this was the case early in the twentieth century with the formation of Ashington Cosmopolitan and Ashington West End, both formed in association with miner-owned and operated social clubs. Additionally, the Clarion Club of Hirst was an affiliate of the Clarion Cycling Clubs, a distinctively working-class organization. Cycling in the coalfield was clearly identified with the miners. However, during the mid-1890s several clubs developed which were located in particular segments of the mining communities.

Between 1891 and 1914 approximately 49 clubs were formed. Twenty-two of these were associated with inns, 14 had no identifiable affiliation and 13 were associated with particular institutions, institutions that fell within the purview of "Rational Recreation" and thus gained the support of particular groups within the community. Clubs developed in association with the temperance movement,

Mechanics' Institutes, churches and youth groups. Vicars, colliery officials and businessmen occupied executive positions in several clubs. These clubs were the backbone of the church parades and parades designed to raise money for charity. What this all suggests is that, like many sports, the membership of cycling clubs did not fall into one particular category but, rather, crossed the spectrum of the inhabitants of the villages. Like-minded groups got together to promote their own interests. Cycling, like the other sports in this chapter, provided leisure activities for a small but enthusiastic segment of the mining communities.

What the history of cycling demonstrated was the fact that the same activity, cycling, found support from different groups within the villages. It was not, like potshare bowling and quoits, rooted in a particular segment of the community. Cycling, as a competitive sport, was closely tied to traditional practices: challenge matches, handicaps, association with inns and innkeepers, and a dependence upon some support from the colliery owners and officials. The development of touring clubs appealed to different groups, groups that espoused the values of "Rational Recreation". It was after the decline of racing that chapels, temperance societies and other self-help institutions became involved in cycling. It was no accident that the church parades began at this time since it was these groups that sponsored and supported the church parades. At the same time, the lines between the different groups were blurred as evidenced by the fact that 45 per cent of the clubs formed after 1891 were associated with inns. Additionally, many of the touring clubs were competitively oriented with rewards for the highest mileage, etc. Cycling provided an avenue for different groups within the villages to come together in social situations. The history of cycling also demonstrated what was to become a greater problem as time progressed: the problems arising out of affiliation with externally controlled organizations, in particular those dominated by Novocastrians. These problems revealed fundamental differences between the cyclists in East Northumberland and the administrators in Newcastle over what amateurism actually meant. For it is apparent that many of the cyclists in the mining villages considered themselves bona fide amateurs. Thus, the ideology underlying the NCU, although the one that dominated throughout Britain, was, in fact, only one approach to amateurism. In other words, the consensus that held amateur sport together was the consensus of a small group of people. Finally, the history of cycling demonstrates vividly the critical importance of the early 1890s. This will be illustrated even more clearly in the history of billiards.

Billiards

For much of the century social life outside the home was limited by the availability of facilities. These were limited to the church and chapel halls, the colliery schoolrooms, local inns and, in some villages, a Mechanics' Institute. Although the activities were developed by groups of miners, in each case the facility was owned and operated by someone else.[39] Except for the inns, all the facilities were heavily dependent upon the colliery owners and officials. There was nowhere that the miners were able to escape the watchful eyes of the dominant groups. This severely circumscribed the activities that could be undertaken. From the late 1880s to 1914, the whole institutional landscape changed in some very fundamental ways. Gradually, miners acquired a greater degree of actual control over these institutions. In a small but subtle change the institutes that developed during the 1890s were Miners' or Working Men's Institutes rather than Mechanics' Institutes. However, the greatest change came in the early years of the twentieth century with the development of miner-owned and operated social clubs. The growth of these clubs, independent of the dominant groups, fundamentally changed social life within the villages. Increasingly, social activities in the evenings and at weekends focused on these new clubs and thus business was drawn away from the chapels and inns. Paralleling these changes was a programmatic shift – a shift from education to amusement and entertainment. At the heart of these changes was the game of billiards.

Sometime during the 1880s, in some Mechanics' Institutes, an apparently small but very significant shift occurred. Their mandate was expanded from a narrowly focused emphasis on education and self-improvement to one that encompassed entertainment.[40] In fact, this apparently small shift represented a major change that opened the door to a massive expansion of activities and membership. This entertainment took the form of a number of games: bagatelle, draughts, dominoes, and, above all, billiards. In 1885, Mr Mark Lambert, owner of Widdrington Colliery, placed a billiard table in a large room at Widdrington Mechanics' Institute.[41] However, it was not until the early 1890s that the game became an obsession with miners across the coalfield. In 1891, Burroughs and Watts, the Newcastle based billiard table manufacturers, either recognizing the potential market in the mining villages or responding to a demand, issued a pamphlet on the rules of billiards.[42] This pamphlet was distributed widely throughout East Northumberland. Whether this was the stimulus or the response is not known but the results were unequivocal – rapid expansion of the game until no village was without its billiard table. In the early 1890s it was possible to identify only four or five institutes playing the game. By 1896, at least 52 Miners', Mechanics' and Working Men's Institutes boasted one or more billiard tables on their premises. Billiards had become a nightly obsession.[43] As early as 1892 every seat in the billiard room at Bedlington Miners' Institute was occupied every night.[44] The managing committee of New Delaval Miners' Institute commented in 1892, with some surprise, that billiards had become the major attraction at the institute during the six months since it was introduced.[45] By 1894 comments were being made about the hold billiards had taken within the mining communities.[46] In Bedlingtonshire, all seven institutes owned tables within two years of the first table being installed in 1892. Subsequently every new institute included a billiard room. One small action, the introduction of billiards, had revolutionized leisure in the villages.

While billiards became an obsession it was, in fact, more important than a simple obsession. The introduction of billiards lay at the heart of significant changes in the social life of the community. In particular, it was billiards that provided the necessary conditions for independence from the mine owners and from the social dominance of the inn. First, the introduction of billiards had a significant impact upon club membership. Several clubs attributed increased membership to the introduction of a billiard table. In 1892, Seaton Burn Mechanics' Institute claimed that "The billiard board which was only laid down in April has proved an immense source of attraction, and has also had the effect of adding largely to the list of members on the books."[47] In 1895, the management committee of Bebside Miners' Institute attributed the increase in membership, from 95 to 200 in one year, to the purchase of a billiard table.[48] Newsham Institute's increase from 40 to 280 in 1898 illustrated the exceptional importance of a billiard table.[49] The consequence of this increased membership was that the institutes became the dominant social institutions in the villages. In 1895, Backworth Institute boasted 750 members and Bedlington 227.[50] By 1897, it was claimed that 60 per cent of the male population were members of Seghill Miners' Institute.[51] Hirst Priestman, in 1903, had 1,273 members, 30 per cent of the male population of Hirst.[52] This was at a time when new social clubs were opening in Hirst. At the outbreak of the war, the 20 clubs in Bedlingtonshire had a total membership of 10,017, over 40 per cent of the male population.[53] This increase in membership and the introduction of billiards provided the foundations for financial stability. Many of the sporting activities within the colliery districts were failures because of financial problems. Billiards overcame this problem and provided significant income for the institutes. A few examples will suffice to illustrate the important financial contributions made by billiards. In 1896, 12 per cent of the income of Backworth Mechanics' Institute was derived from billiards.[54] Newsham Institute, two years later in 1898, earned 61 per cent of its income from use of the billiard tables.[55] Income from billiards in a sample of 16 institutes accounted for 28 per cent of the total income.[56] This often meant the difference between a deficit and a surplus.

What, then, was the role of billiards in these institutes? How extensive was it? And what did it mean to the miners? It was the most popular participant and spectator sport before 1914. It dominated the evenings throughout the year. Nine of the 17 miners talked about billiards at the institute every evening. It cost 1d or 2d a half hour and on Saturdays the line up started at 10 a.m. – a clear indication of its popularity.[57] Although activity sometimes slowed during the summer, it was played in every month. The billiard committees of the institutes organized numerous activities. More than anything, they provided both games and a spectacle for the club members. Although, as we shall see later, inter-club competitions developed, the main focus of activities were the intra-club handicaps, matches, special exhibitions, and tournaments. At the heart of the programme were the numerous handicaps held at frequent intervals throughout the year. Some of these attracted as many as 600 participants while others had a more limited number of 35. Perhaps the importance and popularity of the game are demonstrated most vividly in the case of Bedlington Miners' Institute.[58] During the 1894–95 season, the billiard committee offered a varied programme of 74 handicaps, tournaments and matches for the 227 members. Since many of these events lasted over a few days rarely a day went by without some attraction at the institute. Interspersed among the handicaps were matches pitting individuals or groups against each other, married versus single, for example, and some championship matches. Additionally professional "Cueists" were brought in to give demonstrations. Such was the case in 1906 when Fred Lawrence, the boy champion, gave exhibitions throughout the coalfield.[59] He was followed in 1907 and 1908 by Madam Strebor, the world's lady billiard player.[60] Perhaps even more indicative of the hold of the game was the Bedlington "Strike" of 1897.[61] The Bedlington "Strike" was not over working conditions in the mine but over billiard playing at Bedlington Miners' Institute. Early in 1896, the billiard playing members approached the trustees to ask for structural changes to the billiard room. They requested that a glass roof be placed over the billiard saloon. Twelve months of negotiations with the board of management, also composed of miners, failed to bring agreement. The members then boycotted the institution. Their terrier-like, stubborn resistance won the day. After a three-month boycott, the trustees capitulated and agreed to take a £130 mortgage to effect the changes. The "Strike" had been totally successful. Billiards had become an obsession with the young and the not so young. During the 1890s and the early years of the twentieth century in the 66 institutes and social clubs scattered through the coalfield, hundreds of miners flocked nightly to watch and/or play billiards. The billiard table with its bank of lights, surrounded by seats, became the focal point of night life.

While billiards found fertile ground amongst the miners, it was also unique in that it was not actively suppressed by the colliery officials and mine owners. In fact, on the contrary, it was one of the few activities to be given strong support by the dominant groups, probably because billiards fell within the boundaries of "Rational Recreation" and thus was acceptable to the colliery officials, ministers and teachers. Additionally it was, at first, linked to the Mechanics' Institutes that continued to be strongly supported by the collieries. The nature of this support was illustrated, most clearly, in the opening of Backworth Workmen's Institute in February 1892. The new building had been constructed by the colliery owners, at their own expense. One of the owners, Mr Edward Eccles, JP, donated a new billiard table. R.F. Spence, the colliery manager, and H. Richardson, the agent, were present at the opening which was carried out by Mrs Richardson who then, with Mrs Spence, played the opening game on the billiard table![62] The colliery officials often gave strong support to the development of billiards, a far cry from their approach to most sports. Billiards brought together different segments of the community, albeit in clearly defined roles, and served to reinforce the traditional, deeply rooted relationships that permeated the mining villages.

Intra-club nightly games, handicaps, tournaments and challenge matches filled the calendars of the 66 institutes. Billiards was very much to be participated in. However, billiards did not remain immune from the changes taking place in sport. It soon moved into the world of organized sport where one institute challenged the other. Somewhat surprisingly the miners, renowned for their competitive instincts, did not take to organized competition with a great deal of enthusiasm. Although the first inter-club competition followed quickly on the heels of the explosion of interest in the game in the 1890s, the development of leagues was by no means as swift or successful. In fact, the attempts to form leagues met with a notable lack of success. While 49 of the 66 institutes participated in league competition at one time or another, only seven did so with any degree of regularity – interestingly all located in the northern part of the coalfield. The heart of the coalfield did not, for some reason, take to organized, regular competition. Thus billiards, like many other of the popular miners' sports, was characterized by chronic instability. This was demonstrated in the history of the first league, the East Northumberland Amateur Billiard League (1895–1901). It was inaugurated amidst great enthusiasm in September 1895, when 24 teams from 14 institutes joined the league. Only three of these institutes were members of the league for the six years of its existence. Most of the other attempts to form leagues were even less successful: Bedlingtonshire Billiard League (1897–98) lasted for one year; Hartley and District League (1898–1901) for slightly longer; the Burroughs and Watts Miners' League (1902–07) promoted by commercial interests was no more successful. The East Northumberland Institutes' League, formed in 1907, was still operating at the outbreak of the war. In fact, the popularity of leagues reached its peak in the 1890s; after the turn of the century only 13 institutes entered teams in leagues, less than 20 per cent of the total clubs and institutes. Persistent attempts to promote stable, consistent competition failed. This, of course, was a characteristic of sport in the mining village. Permanence and stability was a characteristic of either middle-class dominated sport or those with a solid commercial foundation.

However, it was within the context of league competition created and run by miners that one of the most interesting insights into mining sport emerged. Nearly without exception, miners played sports and games for money prizes and stakes. It was rare to find the symbolic prizes of amateur athletics in East Northumberland. It was somewhat surprising, therefore, to find that there was deep concern and discussion over amateurism and the definition of an amateur. In billiards, the majority of players were strong adherents of amateurism – they believed, quite vehemently, that they were amateurs. From the outset, unlike the traditional sports, the majority of intra-club competitions were for prizes, not money. Gold and silver medals, cups and even a drilling machine were offered for various handicap events. However, it was the advent of league competition that precipitated the concern over amateur status to centre stage. In fact, their concerns were very similar to the concerns underlying amateur organizations outside of East Northumberland – the concern over equal competition. This was reflected in the definitions adopted at the Annual Meeting of the East Northumberland Amateur Billiard League in September 1897: "Anyone not making his living at billiards is not a professional and is allowed to take part in this competition" and "Anyone who has played for a set stake is hereby debarred and disqualified."[63] Unlike the amateur definitions that dominated amateur sport in Britain, this definition did not contain subjective elements such as "Playing for the love of the game." In other words class elements were noticeable by their absence. The definition was based, more truly than other amateur definitions, on the game being an occupation and the taking of money. What was noticeable by its absence from their approach were definitions such as "Once a professional always a professional" and unknowing contact with a professional being enough to professionalize the offender. However, in East Northumberland, boundaries between amateur and professional were very flexible, the only criteria that applied were those appropriate at the moment. This approach to the question remained in effect during the life of the league and was adopted by the East Northumberland Institutes' League on its formation in 1907.[64] However, problems did emerge as to who was eligible to compete in the league. It is apparent that some institutes imported players for the express purpose of winning the league. This applied particularly in the prestigious Burroughs and Watts miners' competitions. For the first few years the competition was open to any bona fide worker in or about a pit.[65] This was changed in 1901 to anyone on a colliery payroll and a bona fide member of an institute.[66] The definition was changed again, in 1905, to members of an institute who had been members for six months prior to the date of entry.[67] Later, in 1912, a motion was passed to exclude clerks and office hands from the Burroughs and Watts Miners' Championship. What this all reflects is the similarities and differences between the approaches of different groups to amateurism. Like amateur organizations outside the district, miners faced the problems of an over-emphasis upon winning – the importation of players to gain victory – but, unlike amateurs outside the East Northumberland coalfield, the definition was not as restrictive and class based as other definitions. The definition was designed to ensure equal competition.

The history of billiards presented above is very different from any of the other histories. It lay, very clearly, outside the traditional forms of mining sport, in particular, in its association with inns. The above discussion has focused upon the miner-controlled institutes and reflects a distinct departure from the old ways. However, the inns were not entirely absent from the scene. In fact, they were the site of billiards organized along traditional lines, that is challenge matches played for stake money. From the early 1890s, there was evidence of a small number of inns promoting matches and exhibitions. These were limited to two inns in Ashington/Hirst and two in Blyth. As late as 1910, the match between J. Laws, Cambois and C. Dowson, Hirst for £100 was organized and promoted by Henry Bonner of the Market Inn, Blyth.[68] What is clear is that inns and innkeepers played a much smaller role in the development of billiards than they did in many of the other sports.

The massive growth of interest in billiards in the early 1890s led to the development of commercial enterprises interested in promoting the game. First was the influence of Burroughs and Watts, the Newcastle based billiard manufacturers. Their issuing of a pamphlet on the rules has already been mentioned. More important to the promotion of the game, within the coalfield, was their promotion of the Burrough and Watts Miners' Championship in 1894. Reacting to a suggestion by Mr Norman of Backworth Institute, they sent a letter to the secretaries of all the Miners' Institutes in Northumberland, Durham, Cumberland, Westmorland and the North Riding of Yorkshire proposing a competition amongst miners for a silver flagon worth 55 guineas.[69] The response from Durham and Northumberland was positive and thus the first competition took place in October 1894, with the final taking place at Burroughs and Watts Billiard Saloon in Newcastle between the winners of the Durham section and the Northumberland section. Such was the success attendant upon the competition that in 1899 Burroughs and Watts offered separate trophies for Northumberland and Durham.[70] Until the outbreak of the war the finals of the "Burwat" competition became one of the highlights of the year with the winner becoming the recognized Miners' Champion of Northumberland. In fact, Burroughs and Watts did not exhaust commercial interest in promoting the sport. In the two larger towns/villages of Blyth and Ashington–Hirst commercial billiard saloons opened – a sure testimony to the popularity of the game. The Waterloo Road Billiard Saloon and Central Billiard Saloon in Blyth opened for business in 1896.[71] They were followed, in 1899, by the New Billiard Saloon in Station Road, Hirst and by 1902, the Billiard Saloon in Market Place, Ashington. These commercial saloons, along with Burroughs and Watts Saloon in Newcastle, became the site of a steady stream of professional matches. It was out of these saloons that the recognized champions of the coalfield emerged. The first was "Anty" Laws of West Sleekburn who held the title of champion from 1900 to 1908. Laws played regularly for prize money or stakes of between £20 and £50. He was followed by J. Laws of Cambois in 1910. These men were true sport heroes, recognized and celebrated wherever they went.

The emergence of billiards as the most popular sport in the colliery villages represented a fundamental change in the social lives of the villages. No longer were miners restricted to the activities of the churches, chapels and schoolrooms and the "morally uplifting" activities associated with them. Nor were their only alternatives the commercially driven inns. At last they had places where they could be free from the restrictions of officials on the one hand and innkeepers on the other. At last they were relatively free. This was reflected most clearly in the Mechanics' Institutes. The impact of these changes was recognized by the Northern Union of Mechanics' Institutes at their annual meeting held in Blyth in 1898. The discussion focused around the inclusion of billiards in the offerings of Mechanics' Institutes. Several members made comments about the impact of the "great sporting craze" upon the membership. However, after the smoke had cleared only four members spoke in favour of the inclusion of indoor recreation. The majority supported Mr Thompson of Blyth when he stated that "the primary object of the institution was educational, and if they considered the introduction of billiard like pastimes they would be simply turned into a recreation society."[72] They were right – the institutes had, at last, turned into recreation societies and it was this incorporation of education and amusement into the mandate of these institutions that brought fundamental changes to the social life of the colliery villages.

Conclusion

The early 1890s represented a watershed in the social lives of the inhabitants of the mining villages. It was a brief period of time when fundamental changes occurred that had implications for sport, leisure and society. These changes revealed clearly the dividing line between pre-industrial and industrial society. One of the major changes lay in the institutionalization of the social divisions that had always been present in the villages. These were institutionalized in the form of separate clubs for different social groups. The development of clubs with a strong social orientation was illustrative of the fundamental change in how the middle classes approached leisure. In this, the actions paralleled what was developing in the towns and cities across the length and breadth of Britain. While the timing was different, the process of creating socially exclusive clubs was the same. What was different was the fact that this process spread to the villages so that the doctors, teachers and other professionals in Seaton Delaval, New Hartley and Seghill banded together to form socially exclusive clubs.

However, it is in placing these histories in the context of what was happening elsewhere in Britain that some interesting similarities and differences are revealed. Perhaps the greatest differences were in swimming and angling. There was very little similarity between East Northumberland and what was happening elsewhere. Swimming had a long history in Britain.[73] From the 1830s professional swimmers visited many cities. The development of clubs and galas was related to the opening of swimming baths. Thus, in London, the Midlands and Lancashire, mixed clubs were formed and galas held. It was in the 1880s that "amateur" clubs were formed. And it is amateurism that provides a tenuous link with the clubs in East Northumberland. Their development was different because circumstances were different – there were no swimming baths. Thus the competitive swimming was limited to the rivers and most of the clubs engaged principally in sea bathing. Thus emphasis was upon health. As for angling, it had become a mass sport in London, the Midlands, Lancashire, Yorkshire, and especially Sheffield.[74] By the mid-1870s it could truly claim to be a mass sport. Even in Central Scotland, by 1880, 24 clubs had been formed.[75] However, the dominant form of fishing, amongst the working class, was coarse fishing. In East Northumberland a limited number of clubs were formed in the 1880s. They demonstrated two characteristics – limited numbers and fly-fishing! Very different from working-class fishing anywhere else.

With respect to the sports with strong social overtones, their development was somewhat different from the rest of the country. Even though the Newbiggin Golf Club was among the earliest to be formed in England, the history of golf in the district lagged behind the rest of the country. It was not until the early 1900s that the next middle-class club was formed. The Newbiggin Golf Club was different and they were different in their promotion of working men's golf. This initiative preceded anything else in Britain.[76] Why they were different is difficult to determine. Similarly, lawn tennis and lawn bowling were relative latecomers to East Northumberland. By the time the first lawn tennis club was established there, Central Scotland boasted over 20 clubs.[77] Lawn bowling first surfaced in Blyth, a town with a substantial professional and commercial population.[78] In fact, the middle classes of East Northumberland lagged behind their compatriots across Britain, demonstrating, yet again, that East Northumberland was different. Perhaps the most interesting insight lies in the fact that some miners played tennis and even some miners' children were exposed to the game at school.

With respect to the two sports that established deep roots among the miners, it is difficult to assess the degree to which their histories were reflected across Britain. There is no doubt that billiards was widely played but there is no evidence as to the extent, form and role in life. Where it does fit in is in the fact that it developed at a time when good times came to working-class districts. It is, indeed, likely that the experience of East Northumberland may be replicated in the working-class enclaves of Britain. As for cycling, in general terms it fits in with the general history of cycling – starting off in the 1870s with road racing, a shift to racing in enclosed grounds, the development of clubs, the highpoint in the early 1890s, the battle over the "Makers Amateur" question, and the ultimate collapse of racing in the late 1890s. Racing was replaced by the development of touring clubs, with their military type organization, which embarked on rides every weekend throughout the season. What was slightly different was the emergence of church parades that linked them with the landed aristocracy and, thus, the temporary upper-class involvement in cycling. It does appear that cycling took a very strong hold among the inhabitants of East Northumberland. In the period 1876 to 1900, 65 clubs graced the roads of the county. At the same time in Central Scotland, with a larger population, only 46 clubs were formed – a significant difference.[79]

6 Football

A symbol of sport in East Northumberland[1]

If there was one sport that brought together the past, present and future, it was football. Deeply rooted in tradition, it became the *sine qua non* of modern organized sport. More importantly, its history reveals the complexity of its development. In fact, histories of football which have focused on the permanent, high profile professional teams and leagues have served, I suggest, to obscure the real history of football, its many layers, and the ways in which the past and the future existed together.[2] Even in 1914, football was a complex amalgam of traditional practices and the modern game. Its history reveals the significant changes that took place over a 30 year period. The history of football and its relationship to life in the villages will be explored, first, by examining its roots in traditional practices. This leads to an examination of the changes in the game that took place during the period up to 1914. This provides a framework for a discussion of the various layers of football and the reasons for the success and/or failure of clubs. We will then examine the ways in which it penetrated all corners of society. Finally, the history of football will be placed within the context of changes in village life.

Football was not a stranger to the miners of East Northumberland. As early as 1634 football was forbidden in the area. There are references to the game being played in the 1820s. What is unknown is the form it took. Like potshare bowling the rules were never written down but were passed from father to son. Our view of the game becomes clearer in the second half of the century. By the 1860s, there were generally accepted rules because infrequent, but regular, games were played during the summer months. These games were held in association with the annual village festivals. During the summer of 1860 football tournaments were held at the 5th Annual Great Northumberland Temperance Festival at Cramlington Colliery, at a gala at Scotland Gate in August, and, in September, at a temperance festival at Seaton Delaval.[3] Unlike the modern game, no teams were announced and, in fact, it is impossible to know what form the game took. The popularity of the game was demonstrated, in 1866, when it was included on the programme of the third Annual Miners' Picnic on Blyth Links. From then on most of the Miners' Picnics included the game in the day's festivities. In the ensuing years flower shows, schools, cooperative societies, the International Order of Foresters and other organizations included the game in their annual programmes. Thus, football, with a generally accepted set of rules, was well established in the mining communities well before association football took hold in the early 1880s.

The stimulus to the development of the new form of the old game, "Association Football", came from outside the district – Newcastle. The association game was first played in the North East in January 1875.[4] Two years later it was estimated that there were only 250 players on Tyneside. By the early 1880s, it was relatively well established on the banks of the Tyne with over 1,800 active players. While there was some football played in Blyth and Bedlington in the early 1880s, the first unequivocal evidence of miners playing the game occurred in October 1882 in Bedlington, when a team of tradesmen played a team of miners.[5] This was the beginning of what was a virtual explosion of interest in this new game. In 1883, there were only five teams in the coalfield, During the next two years, the number of teams increased dramatically and by 1885 there were 62 teams from 48 clubs located in 20 villages and towns.[6] The popularity of the game and the characteristics of competition were reflected in the first real focal point of the game. Between 1884 and 1886, Shankhouse, population 1,000, boasted 17 teams – teams with names such as Royal Blues, True Blues, Albion, Red Rose, Successors, and Water Dogs. Only Shankhouse Black Watch and Shankhouse Red Caps saw a first anniversary.[7] Football had become an obsession with the male population of the village. The example of Shankhouse was typical of the state of football at the time, a large number of impermanent teams playing a variety of challenge games, mainly in the summer months. For example, between June 19 and September 11, 1886, Shankhouse Water Dogs played in five tournaments associated with annual picnics – Oddfellows' Picnic at Horton, Cramlington Sports, the Foresters' Galas at Dudley and Cramlington and the 13th Annual Ashington and Bothal Flower Show.[8] In fact, this typified the dominant form of football until the First World War. As late as 1913 flower shows were still including football competitions in their offerings, challenge matches were still being played, and groups were still organizing teams for special occasions, such as when Levison's Pilgrims played the Spion Kop Jumpers at New Delaval in a benefit match for two sick workmen.[9] More importantly, there was always a greater number of these impermanent teams playing in a variety of competitions, exhibition games and challenge matches than there were permanent teams playing in leagues. Yet it was these semi-permanent teams playing in organized competition that came to dominate, in the public's mind, the football scene and brought them within the not-so-friendly embrace of the middle-class and Newcastle-dominated Northumberland Football Association (NFA).[10] This was important because it was the NFA sponsored competitions, in conjunction with the FA Cup, that provided a degree of structure to the year. Additionally, by the turn of the century, any teams aspiring to championship competition had to affiliate with the NFA. By 1887, a limited number of clubs from the colliery villages were playing in the annual NFA Senior, Second Team and Junior Cups. It was the growing success of the coalfield-based teams in these competitions that precipitated the first battle against the Newcastle dominated NFA.

In April 1888, 18 clubs in East Northumberland, because of perceived Newcastle domination of the NFA, established their own organization, the Tyne and Tweed Association.[11] In 1889, the renamed Northern FA applied for recognition from the

Football Association (FA.).[12] The FA refused to recognize the Northern FA.[13] The association collapsed and was forced within the control of the Newcastle-dominated NFA. The dye was cast. Coalfield clubs, interested in pursuing the highest level of amateur football, were forced within the unfriendly confines of the NFA. The disagreements between the two groups remained one of the defining characteristics of football in Northumberland until the war.

The period 1890 to 1898 witnessed great growth in the number of coalfield-based teams competing in NFA competitions – from 11 in 1891 to 38 in 1898. At the same time we witness the gradual development of the competitive structure which came to characterize football. At first, competitions were carried on within the traditional practices, challenge matches and cup and medal competitions sponsored by innkeepers, football clubs and charitable organizations. During the early 1900s, a number of cup competitions were inaugurated with widespread support from the dominant groups. These competitions raised funds for charities and, in particular, hospitals. The two most prestigious, attracting teams from the whole of the North East, were the Tynemouth Infirmary Cup (1904) and the Royal Infirmary Cup (1906). At the same time, there were four competitions for teams in East Northumberland. There were two other local competitions that gained widespread support in the villages: the Henderson Cup (1895) and the Aged Miners' Home Cup (1908). These, in addition to the NFA competitions, provided the highlights of the season.

However, the most significant change was the institutionalization of the "football season" – a season from September to April. For a small group of the more permanent and high profile teams a clear structure was placed on the football season. In 1889, Shankhouse Black Watch joined the short-lived North Eastern Counties League which attracted teams from as far south as Redcar. This league collapsed after one year but it was followed, in 1891, by the formation of the Northern Alliance (NA). Shankhouse and other district teams participated in this league which remained one of the premier leagues until the war. More important to the development of football in the district was the inauguration of intra-coalfield leagues. Between 1894 and 1898 three relatively permanent leagues were formed: the East Northumberland League (ENL) (1894–1910); the Wansbeck Junior League (WJL) (1897–); and the East Northumberland Combination (ENC) (1898–1910). Between 1890 and 1898 the football calendar became institutionalized. For example, in 1892, 30 per cent of Shankhouse Black Watch's games were exhibition games. By 1895 the number of these games had diminished to 15 per cent. In 1897, Ashington's lengthy 53 game schedule included no friendly or exhibition games. By 1898 clubs arranged fixtures before the season, not, as had been the practice, during the season. Modern sport had arrived.[14]

The institutional structure that was in place by 1898 remained intact until the war. The first decade of the new century witnessed the expansion of relatively permanent football clubs into nearly all the villages in the coalfield. The number of mining teams competing in NFA competitions rose from 21 in 1900 to 63 in 1908. This was paralleled by the development of locally based leagues: the Blyth and District League (BDL) in 1904 and the Ashington and District League (ADL) in 1911. The other significant change occurred in 1913 and 1914: professional football finally gained a foothold in the area with the acceptance of Blyth Spartans (1913) and Ashington (1914) into the North Eastern League, a professional league formed in 1907.

By the outbreak of the war, football had taken deep root in East Northumberland. Most villages boasted more than one team and the majority contained enclosed football grounds. Throughout the football season, September to April, every Saturday witnessed numerous contests in the coalfield. Football appeared to be in a stable condition with a county organization, the NFA, overseeing the highlight of the season – the various cup competitions. Additionally there were leagues within the coalfield and ones that pitted the colliery villages against other teams in Northumberland and Durham. At the pinnacle lay Newcastle United, the standard bearer of the hopes and aspirations of the North East. While it was undoubtedly true that football played a significant part in the lives of most males in the district, this picture of stability and permanency was, in fact, an illusion. Football was beset by systemic instability and a chronic lack of permanency. The edifice of football was based on a foundation of continuing change. The majority of clubs, those lying outside the jurisdiction of the NFA, rarely existed for more than a year, yet they were replaced, immediately, by other teams which led just as brief lives. Even within the NFA, the site of organized competition, there was a pronounced lack of permanency. Of the 408 clubs from East Northumberland that participated in the annual NFA Cup competitions between 1890 and 1913, only 15 (4 per cent) competed for 10 or more consecutive years.[15] Lack of permanency was one of the defining characteristics of football and, indeed, all sport in East Northumberland.

Although the lack of stable, permanent clubs was a basic characteristic of the game, there were differences that serve to focus on the causes of this systemic instability. Every year there were more teams playing beyond the jurisdiction of the NFA and the leagues than within them.[16] Created by groups interested in playing the game, they never developed the organizational infrastructure that was necessary for success in the new game. It was in the NFA cup competitions and leagues that a degree of stability was provided. Even then only 137, exactly 50 per cent, of the 274 clubs that entered the NFA competitions did so for more than two years. However, there were differences amongst the NFA clubs. Forty one percent of the senior clubs played in cup competitions for five or more years, while only 26 per cent of the minor clubs and 8 per cent of the junior clubs did. Stability was, in part, a function of level. This instability was also reflected in the history of the various leagues. Of the 19 intra-district leagues that were formed between 1894 and 1913, 12 lasted one or two years. Even the semi-permanent leagues were plagued by instability. The most important leagues, the ENL (1894–1910) and the BDL (1904–14), had no teams that played throughout their histories. In fact, 17 (31 per cent) of the 54 teams that played in the ENL only lasted for one year. Similarly, 19 (33 per cent) of the 57 that played in the BDL lasted for one year. Seventy-five per cent of the ENL teams and 83 per cent of the BDL teams played for less than 5 years. This instability was even greater in the junior ranks:

45 per cent of the 60 teams playing in the Wansbeck Junior League (1897–1914) succumbed after one year and 94 per cent were gone after four years. There was very little continuity in organized football. At the same time, the number of teams increased over the years as the *game* of football became more deeply rooted in the villages.[17]

Instability had always been a defining characteristic of sport in the villages. There were few institutions or clubs with any degree of permanency. Characteristically the most popular sports of potshare bowling, quoits and handball were rooted amongst the individual working miners. The stimulus to competition came from the grass roots and was based in the challenge system. Thus, the regularity of matches for individual miners depended upon the number of people who wished to challenge or be challenged. The early years of football faithfully adhered to these practices, teams were formed for particular occasions, focused on annual celebrations, played on any available ground, and were rooted in the challenge system. It was the development of leagues in the 1890s that precipitated football into the world of modern sport with its new requirements. The emergence of regular schedules of games placed new demands on the emerging clubs – an enclosed ground, money for travel and, soon, money to pay players. These necessary conditions were a prerequisite for *all* teams that wished to succeed in league football, not just for the few who strove to reach the top. To survive, all league teams, no matter what the level, had to become, to some extent, commercial enterprises. The traditional supporters of miners' games, inns, landowners and colliery managers, gave some support. However, it was soon apparent that this was insufficient to meet the demands; the vast majority of teams failed. New conditions developed that became necessary for the success of football clubs in the district. The conditions underlying the failure of so many teams and the conditions necessary for permanence can be illustrated through three case studies. First, the reasons why so many clubs failed are clear and will be examined through the histories of several minor teams. Second, these conditions, when added to a failure to adapt to new conditions and the changing face of football itself, led to the demise of even successful clubs. The history of Shankhouse Black Watch, the most successful team in the nineteenth century, demonstrates the necessary conditions for success and the reasons for failure. Third, Bedlington United, on the other hand, demonstrates the necessary steps that had to be taken by a minor club to raise itself to a higher level.

While it is important not to oversimplify the causes underlying the chronic instability of clubs it is possible to identify the two most important factors that led to the demise of so many clubs – financial viability and access to a playing field.[18] Although both, independently, were sufficient cause for the termination of a club, they were intricately related, in that an enclosed ground was the *sine qua non* for the charging of gate money, the biggest single source of revenue for the clubs. As a result of forces that lay beyond their direct control, many clubs led a precarious existence teetering on the brink of disaster from year to year. Withdrawal of permission to use a field spelt disaster. This was especially the case for junior clubs "which were wont to play on any waste land."[19] Strikes, lockouts, and depression in the coal trade always brought a serious diminution in gate money. The foundations for success were recognized as early as 1893 when the decline of football in Morpeth was related to the problems of expenses, the decline in gate money and the need for alternative sources of money.[20] In 1896, Blyth FC was plunged into a financial crisis as a result of slack trade which had caused several players to leave the district in search of work.[21] The bad times of the early 1900s, that brought major problems to leagues, teams and competitions throughout East Northumberland, were attributed to the "chronic poverty" that beset them all.[22] Several teams collapsed because they lost their grounds. Others were forced to disband because of financial problems. Blyth Swifts, after a 17-year history, during which they frequently fielded three teams, were forced to disband in 1909 as a result of financial difficulties.[23] Perhaps the precarious position of minor and junior clubs is best illustrated in the case of New Delaval Star, which withdrew from the Ashington and District League in April 1913.[24] The reason given for their withdrawal was a financial deficit. This was due to the lack of support and thus of gate money. During the season the gate receipts totalled £3 and 15 shillings. This was augmented by £1 and 3 shillings from the players' pockets and £6 and ten shillings raised in a lottery. The income was insufficient to cover costs – the field, rented from a local coal owner, 6 shillings; the rent for the "stripping" quarters, 15 shillings; referee's fees of 4 shillings a game; and finally repairs to the ball! The margin between success and failure was thin and, thus, the existence of clubs and leagues was heavily dependent upon conditions over which they had little control – the health of the coal trade, decisions of land and colliery owners, access

Plate 6.1 Barrington Villa, early 1900s.
Reprinted with permission from Martin, E. (1997) *Images of England: Bedlingtonshire*, Stroud: Tempus Publishing

Plate 6.2 A Barrington team just before 1914.
Reprinted with permission from Martin, E. (1997) *Images of England: Bedlingtonshire*, Stroud: Tempus Publishing

to an enclosed field, and the success of a variety of fund raising activities. In fact, financial viability was achieved only under certain circumstances. The delicate balance between success and failure and the necessary conditions for survival are illustrated in the history of Shankhouse Black Watch, the first powerhouse in the world of East Northumberland football.

Formed in 1884 in the small village of Shankhouse, Shankhouse Black Watch quickly moved to a position of pre-eminence in the world of local football. Universally recognized for their contributions to the game in the district, the Black Watch, a team of local amateurs, dominated senior football until the late 1890s. Winners of the NFA Senior Challenge Cup, the symbol of football supremacy in Northumberland, on six occasions between 1884 and 1895, they also won the Tyne Charity Shield on several occasions. In 1887, the Black Watch reached the Fifth Round of the FA Cup competition, eventually losing to Aston Villa, 4–0, at St James' Park, home of West End Football Club, which became Newcastle United in 1892. They joined the Northern Alliance on its formation in 1891 and won the Championship Medals in their first season. During the 1890s they always challenged for the championship of the NA. Without a doubt, Shankhouse Black Watch was the dominating club in senior football in the North East. By December 1905, the club was on the verge of disbanding and was saved, temporarily, by funds raised at a public meeting held at the Albion Inn, Shankhouse in December. Ending up bottom of the NA in March 1906, they failed to gain re-election to the NA –

the great days were over. Dissolution was avoided by amalgamating with Shankhouse Albion. What, then, accounts for the rise and fall of Shankhouse?[25] The reality was that the world of football had changed to such a degree that it was virtually impossible for a team such as Shankhouse to succeed in the rarified atmosphere of senior football – time had passed them by.

Throughout the first 12 years of its existence, Shankhouse Black Watch was an amateur team consisting of local players who played their games on ground rented from the owner of Stickley Farm. Much was made of this fact and even as late as 1895 the newspapers reported that the team comprised local players and was "strictly amateur".[26] In fact, in 1891, at least seven of the players were coal miners.[27] By 1896, Blyth, their opponents in the NA, admitted to paying players to keep them in Blyth when there was a downturn in trade.[28] By 1897, professionalism was infecting even the less prestigious East Northumberland Amateur League. At its annual meeting in June 1897 the delegates agreed that all teams paid players.[29] From the mid-1890s many teams in the coalfield either paid the players or provided work, they certainly were not truly amateur. Additionally, by the mid-1890s, teams in all leagues were importing players. Payment was an accepted part of football. Thus, increasingly, Shankhouse faced professionals on the field of play. By 1900 they participated in the NA with the professionals on Sunderland A, Middlesborough A, and Newcastle A, a team that included Wilf Innerd, later captain of Crystal Palace. It is doubtful if Shankhouse maintained its purely amateur approach much beyond 1896, but it could not compete, financially, with many of the other teams. This brings us to the Achilles' heel of all football clubs in the district, financial viability.

The success of Shankhouse on the field was predicated on having sufficient finances. Money was raised in various ways: membership schemes, donations from prominent people in the locality, the running of bazaars, football medal competitions and annual sports. However, gate money was the most important source. This was related to the ability of the team to attract spectators. During the 1880s, the Black Watch regularly attracted spectators from all over the district. They were clearly the best team in the coalfield. With the expansion of the number and rise in calibre of teams in the 1890s, Shankhouse's pre-eminent position was threatened. At the same time, entrance into the NA significantly increased travel expenses. Additionally they undertook expensive ground improvements. Arcot Park was opened in 1896 with two dressing rooms, a press box, and a fully enclosed field.[30] At the same time, executives began to comment on the declining gate money. In 1898, for the first time, the club reported a £14 loss and questions were raised about the future of the club.[31] In reality, this was the beginning of the end. While they commenced a number of schemes for fund raising, none met their needs on an ongoing basis. However, gate money continued to decline. A club with limited resources could not compete for players with the other teams that emerged to take over the role of premier team in the district. In particular teams in Ashington and Blyth moved into the NA. Shankhouse became less and less successful on the playing field. In 1892 they won 79 per cent of their 34 games. By 1902 this dropped to 41 per cent of 32 games, and worse was to come. In the 1905–06 season, after

11 games they were winless. Just as important to the declining fortunes of the club was the entry of Newcastle United into the First Division in 1898. St James' Park became the mecca for football enthusiasts in the villages. The time of Shankhouse had passed. Gate money continued to decline. By December 1905, the club was in a financial crisis. The club was not re-elected to the NA in 1906 and was only saved by amalgamation with the other village club, Shankhouse Albion. The newly named Shankhouse FC joined the BDL in 1906, a league more appropriate to a village the size of Shankhouse. They immediately became a force to be reckoned with in the league, winning the championship in 1908. At the same time, the financial status of the club improved and by 1908 they posted a balance of over £16. Shankhouse continued to harbour visions of moving back to what they considered their rightful place amongst the premier clubs of the district. Annually they played in the NFA Senior Challenge cup, all to no avail. Additionally, every June at the annual meeting of the NA, they applied for re-entry to the league. Every year they were turned down. The once great club had simply become one of the minor teams in the villages of East Northumberland.

Shankhouse was, in many ways, atypical. Few clubs lasted for more than two or three years. The majority failed to provide the infrastructure of support and maintain financial viability. A relatively permanent infrastructure and financial viability required certain basic conditions. These are illustrated in the history of Bedlington United. United was formed in 1897 in the same way that the majority of clubs were formed – by a group of interested individuals who, for the first few years, supported the club out of their own pockets. From the outset the Bedlingtonians had higher aspirations. In 1900, they joined the ENC. In 1901 they entered a team in the NFA Junior Cup Competition and in the following year entered teams in the NL, the WJL and the NFA minor and junior competitions. Like many clubs their aspirations had advanced beyond their resources and, in May 1903, United reported a small deficit. Immediate steps were taken to reduce their expenses and debt. They decided to concentrate their efforts on one senior team. On October 31, 1903 a bazaar was held to clear off the small debt and put the club on a sound financial basis.[32] The deficit was eliminated. One of the key figures in bringing about this change was Dr R. Morris of Bedlington, and it was this association with the social elite of the town that became one of the key variables in establishing financial security. During the next three years they established themselves as one of the pre-eminent clubs in the district. By early 1906, they, along with Blyth, had distanced themselves from other ENL clubs, both on the playing field and at the turnstile. Bedlington looked for greener pastures – a place in the NA. This shift in status was reflected in their move from the NFA minor competition to the senior competition. Now they played with the premier senior teams in Northumberland – Ashington, Blyth Spartans, and Morpeth Harriers.

1906 was a pivotal year in the fortunes of Bedlington United. It was the first year in which a reasonable balance, £41, was left after expenses had been met.[33] The income of £149 was derived, in the main, from gate receipts (64 per cent). This was the beginning of the move from being a team, like the majority of other teams, which lived on the knife-edge of disaster, to one that became a profitable enterprise.

Income increased steadily over the years reaching a highpoint of £669 in 1913. This left a very healthy balance of £226. This increase, however, was not due solely to the increase in gate money, which was only 40 per cent of the income in 1913, but to a new source of income, transfer fees. Transfer fees were first reported in 1907 (£35).[34] These rose to £280 (42 per cent of income) by 1913. In fact, they accounted for a greater proportion of the income than did gate money. Just as important was the fact that it was becoming a profitable business enterprise, as the expenses as a percentage of income declined from 78 per cent in 1907 to 66 per cent in 1913. This shift towards being a profitable business was paralleled by increased professionalism in their approach to players and to considering the club as a business. In fact, 1907 was the last year that the club straddled the thin line between amateurism and professionalism. The players only received money in lieu for loss of work and travel. This was removed in 1908, when they were successful in their application to the NA. Bedlington United had now moved into the top flight of Northumbrian football and played in the NA, the NFA Senior Cup, the FA Cup, and, after 1911, the senior section of the Aged Miners' Cup Competition. By 1909, the players were receiving money "[w]ell" above the county average of a miner."[35] From thenceforth questions of remuneration of players were raised at the annual meetings held at a Bedlington inn. Additionally, moves were made to improve the facilities at their Hollymount ground. While the suggestion to build a grandstand was defeated in 1910, it was passed, in 1913, that a grandstand be constructed at a cost of £100.[36] It was built with considerable financial support from Mrs Burdon of Hartford Hall. Bedlington had become a profitable business enterprise. The revenue-generating potential of football was recognized by the Bedlington committee when they rented out advertising space at the ground. Strong support from the community, a group of competent administrators, and a substantial financial return were the necessary ingredients for success.

It was the payment of players in the NA and even the ENL that brought major changes. The payment of players was not limited to the openly professional clubs but soon permeated all teams in the district. This brought the district clubs into conflict with the NFA, the Newcastle controlled body that lay at the heart of the development of organized football in Northumberland. Payment of players was anathema to the NFA. This stimulated under-the-table payments to players. The problems of these payments and the poaching of players escalated. At the 1895 annual meeting of the NFA some of the clubs from East Northumberland moved that clubs be allowed to pay players in order to prevent poaching; the motion failed.[37] However, the problem did not go away and clubs were confronted with the dilemma of either being truthful and recognizing that the payment of players was rampant, or being hypocritical. At the annual meeting of the East Northumberland Amateur Football League in June 1897, after extensive discussion, they decided to eliminate the word "amateur" from the title of the league. Basically everyone agreed that all teams paid players and that "if the East Northumberland clubs are to be honest we must strike out the word amateur."[38] This ran counter to the NFA who, only the year before, had disqualified Blyth from NFA competition because they had formally decided to operate the club on a professional basis. In

1899, two ENL clubs, Burradon and Ashington, made the same decision. However, they continued to play in the NFA Senior Cup – the NFA had changed its position. From thenceforth, teams from the NA, ENL and BDL, all openly paying players, played in NFA competitions. Amateur clubs were noticeable by their absence from the coalfield.[39] Payment of money either for lost time or as payment to play was a basic ingredient of football in East Northumberland. This placed them beyond the pale of amateur football.

The growing popularity of football during the last two decades of the century represented a leisure revolution. Within the space of two decades the focus of Saturday afternoons shifted from the Town Moor, Newbiggin Moor and the beaches to the fields and stadiums located in the villages. This shift was more than a shift in leisure-time activities and represented some fundamental changes for life in the villages. Football was different from potshare bowling, quoits, pigeon flying and the other traditional sports – it was a team sport which required a field to play on, a ball to play with, and an organizational infrastructure to bring some coherence to the game. It was the forerunner of the modern sports that revolutionized life in the villages. No matter how temporary the team or low the level of football, they required a field to play on and it was the development of playing fields that revealed some subtle but fundamental changes that were taking place within the villages. The necessary conditions for playing football illustrated some basic changes in the social relationships that altered the traditional power relationships within the villages. They also changed the physical structures of the villages, a football field became a fixture in many of the villages and provided a new focal point of community activity and pride. The development of these grounds provides new insights into the nature of the communities.

From the outset, football was dependent upon the support of friendly farmers, colliery owners, colliery officials and the major landowners. In the 1880s, for the mass of transitory teams that flitted across the playing fields, the fields must have been rudimentary indeed – the legendary cow pasture. There was no permanence, permission being granted for the particular occasion. This type of arrangement remained the basic staple for football until the war. It was amongst teams with a degree of permanence that it was possible to discern the processes underlying the development of relatively permanent enclosed football grounds. All the grounds were developed for teams that participated in leagues and the NFA Cup competitions. However, even amongst these the number that achieved any degree of permanency was relatively small. For example, of the 169 clubs that entered teams in the NFA Minor Cup competition between 1891 and 1914, only 20 (12 per cent) lasted for longer than 5 years and it was these clubs that developed grounds. Additionally it was this group of clubs that were the only relatively permanent members of the ENL/ENC and the BDL. These, in addition to the limited number of senior clubs who played in the NFA Senior Cup competition and played, at some time, in the NA, were the only ones that developed the permanent grounds. In fact, it is possible to identify, quite accurately, the pattern of the development of enclosed football grounds. Prior to 1893 there were four or five enclosed football grounds, all multi-purpose grounds. Between 1893 and 1898, at least, 30 new grounds were opened. This paralleled the development of the ENL and the WJL. After that there was steady growth to 1914 when it was possible to identify 52 grounds. The question is, what were the forces underlying the development of the grounds?

The development of enclosed football grounds in the 1890s brought into play a complex of forces that illustrate the nature of the interactions between the different groups within the community that underlay any decisions with respect to land usage. At the heart of the process of negotiations over land usage lay the land and colliery owners. Even though their absolute power was undermined by the Local Government Acts of 1888 and 1894, they were still central to the process and their permission was still a necessary ingredient, as was illustrated in the unsuccessful attempts of Hirst United to get a ground in 1906. Several deputations had tried their level best to secure a ground but had failed. The Duke of Portland's agent (Mr Sample of Bothal) refused to sanction the local farmers to let any land. They had deputations to the Ashington Colliery Offices asking the directors and the owners of the North Seaton estate to let them a piece of ground. This permission was refused because the ground was needed for building purposes.[40] Thus, the colliery and landowners presented a barrier to any attempts to develop grounds.

More important to the actual creation of grounds were the on-site colliery officials and local farmers, on the one hand, and impetus from within the communities on the other. As early as 1886, J.G. Weeks of the Bedlington Coal Company rented a field to Bedlington Coop Football Club. In 1893 the owners of Seaton Delaval Colliery, through their agent R.E. Ornsby, provided a ground for Seaton Delaval F.C.[41] Seaton Burn Wanderers, in 1895, played on a field lent by Mr J.W. Rutherford of North East Mason Farm.[42] These examples can be replicated for any year of the 1890s. However, more important to the development of more permanent enclosed grounds were a number of initiatives arising from within the communities. Several of the enclosed grounds were associated with community recreation grounds created by the Urban District Councils, on the one hand, and community groups on the other. In 1900 and 1909, Earsdon Urban District Council gave permission for teams to play on Holywell Recreation Ground and Shiremoor Recreation Ground.[43] Arising more directly from within the mining communities themselves was Bedlington Station Recreation Ground in 1893, a ground owned and operated by miners associated with Bedlington YMCA. In 1896, miners associated with Seghill Workmen's Institute took over six acres and laid them out as a football and cricket ground.[44] They fenced the football ground in 1899. These multi-purpose grounds were the focal point of many of the teams in the ENL and ENC. It was at the highest level of football that custom-made grounds emerged.

Appropriately, one of the first custom-made grounds to be developed was associated with the first powerhouse of East Northumbrian football, Shankhouse. In 1896, Arcot Park was opened, by Shankhouse FC.[45] In the following years grounds sprang up throughout the coalfield: New Delaval, 1896; Choppington, 1897; and Seaton Delaval's "splendidly appointed" Moor Edge Ground in 1906.[46] Throughout the mining villages these football grounds became the focal point of the sporting life of the village during the football season. This represented a significant shift in focus away from the previously pre-eminent inns. However, it was the successful teams in the Northern Alliance that created the most ornate and successful football grounds. In 1908, Ashington FC terminated its 20 year association with the Ashington Recreation Company and moved to the newly erected Portland Park, a ground built specifically for football. Portland Park with seating, changing accomodation and offices was so successful that they embarked upon building a covered stand in 1912. Similarly, Blyth Spartans opened Croft Park in 1909. Finally, in 1913 Bedlington United added a covered stand to their ground at Hollymount. All three clubs played at the highest level of local football.

Unlike many sports, football cut through the divisions within the villages and was adopted, to some degree, by all segments of the community. By the turn of the century, few people remained untouched by the football mania. At the same time the nature of its development – who promoted the game, who organized the game, and who played the game – revealed some significant changes. In many respects, it maintained its connection with the traditional practices that undergirded sport, in particular with respect to its attachment to the inns and innkeepers.[47] The first club within the district, Bedlington Burdon, held its inaugural meeting at an inn in Bedlington in July 1883. During the ensuing 30 years the inns maintained a role in the football world. Many clubs were formed at particular inns and maintained an attachment to them. In fact, over 70 per cent of all club and league meetings, between 1890 and 1914, were held at inns.[48] As with other sports, the innkeepers were active in promoting the game. For a brief period in the 1890s, during the transition from the old forms of competition to the new, a number of innkeepers promoted competitions in fields adjacent to their inns. This brought them into conflict with the NFA who, in 1896, refused to grant permission for the competitions to be held. Inns remained central to the development of the game and innkeepers became active on the executive committees of the most prominent teams, Blyth Spartans and Ashington FC.

While traditions maintained a robust presence, football spread beyond the old boundaries into other corners of mining society. Providing a clear alternative to the dominance of the inns were the new Miners' and Working Men's Institutes that developed during the 1890s. Very soon miners used these institutes, and after 1899, the social clubs, to organize football clubs. Thus instead of the friendly "pub" the institutes became the site for weekly, monthly and annual meetings. From 1895, many clubs became affiliated with or used the Miners' Institutes. Football reflected the significant changes taking place within the mining villages, the development of institutions initiated, organized and administered by miners.

While the shift from the inns to the institutes was important as an illustration of the growing independence of the miners, it did not reflect a major shift in terms of what was perceived to be important. Far more important, in terms of real change, was the adoption of football by schools and churches. Although there were church-based teams from the mid-1880s, it was not until the mid-1890s that concern was expressed in several quarters over the impact of football on church and Sunday School attendance. There were, however, differences between the denominations

Plate 6.3 St Stephen's Church, Seaton Delaval. Bible class football team of 1910/11. Reprinted with permission from Allan, T. (1990) *Seaton Delaval: A History in Photographs*, Morpeth: Northumberland County Library

in their approach to football. From 1895, Anglican churches and missions promoted clubs in association with their congregations. For example, Seghill Church Rovers were formed at a meeting convened by Rev. G.H. Holoran, curate at Seghill, in February 1898.[49] By the turn of the century, Anglican churches and missions, throughout the coalfield, promoted football for their parishioners. Not only did the Anglican churches promote teams, but by the early years of the century they engaged in league competition. In September 1913, four Anglican churches joined with Hirst Free Church to form Seaton Hirst and District Church League.[50] By the outbreak of the war, 14 Anglican churches and missions were participating in regular competition. Football had become a recognized element of church life and activities. For the most part, football was never adopted by the more important and pervasive Methodist chapels. At the same time other denominations did flirt with football. Limited numbers of Presbyterian, Catholic and Primitive Methodist churches/chapels did promote the game in the 1900s. Unlike the Anglicans, Methodists, Presbyterians and others remained lukewarm in their approach to football.

Far more important, as an indication of the degree to which football penetrated society, was the acceptance of the game in the schools. When schools actively promote something it means that it has an important place within society and, in particular, within the ideology of the dominant groups. In fact, one or two school teams played prior to the 1890s. However, it was in the 1890s that there was an explosion of interest in the game. The first clear, unequivocal evidence of elementary schools within the colliery villages playing football was their entry into the Sambotee Cup, a trophy presented by the Sambotee Tea Company, in 1894 and 1895.[51] In 1895, nine schools from Blyth and the vicinity entered the Blyth Schools' League.[52] This was the beginning of significant expansion of inter-school football within the district. On October 11, 1902 the East Northumberland Schools' League was formed.[53] This became a regular fixture within the coalfield until the war. As football gained increasing popularity, the coalfield divided into districts with locally focused leagues. In 1905, the 10 school Seaton Delaval and District League commenced operations.[54] Ashington and Hirst Schools played on a regular basis from 1902. 1912 witnessed the formation of South Northumberland School League. The importance and popularity of the game were reflected in the donation of cups, trophies and shields for competition amongst the schools by local dignitaries and institutions.

As the popularity of football increased, moves were made to expand competition beyond the boundaries of the coalfield. In November 1897, an association was formed to compete for the Northumberland and Durham Schools' Cup.[55] In 1898 Shankhouse British School became the first school in East Northumberland to enter the competition. With the formation of county organizations there developed inter-county or district competitions between teams chosen from the schools in a district. The first team to represent East Northumberland was chosen in 1903. The popularity of these games was such that one team did not suffice. In 1906 the district was divided into two regions. By 1910, they had subdivided into five districts, each with representative teams that played against the other districts in Northumberland and Durham and, after 1907 and the inauguration of the English Schools' Cup, in the national competition. These actions culminated in the formation of the Northumberland Schools' Football Association in 1911. By 1913 the association boasted 62 teams, 20 of which were located in the mining districts.[56] Football had become an important element in the educational system; it had become institutionalized.

The changes in the leisure practices of the miners, as reflected in the history of football, were nothing short of revolutionary. Between 1882 with the formation of the first football club and 1914, the whole landscape of leisure activities was changed in fundamental ways. More than that, however, the changes reflected more fundamental changes in the whole culture of the mining villages. Not only did leisure practices change but daily life within the individual villages changed. The history of football illustrates, first, the expanding horizons for the inhabitants of the villages both within East Northumberland and outside the district. Second, the way in which they played football provides some glimpses into the nature of mining society and how it differed from the rest of the world. Finally, the meaning of football to the miners and the mining villages provides some unique insights into the nature of the changes that swept through the coalfield in the 30 years prior to the First World War.

Prior to the 1880s, the sports of the miners were played within the friendly climes of mining society. The majority of competitions in potshare bowling, quoits and pedestrianism were against other miners. These, for the most part, involved individuals from within the same villages. It was only at the higher level of competition that miners from one village competed against miners from another village. The world of individual miners was, for the most part, severely circumscribed. Football changed all that. During the 1880s and early 1890s most competitions were against teams from within the immediate vicinity. It was the development of league competition that brought teams from throughout the coalfield into regular competition. The formation of the ENL in 1894 ushered in a period of expanding opportunities. Every weekend teams travelled throughout East Northumberland visiting other colliery villages. The horizons of individual miners expanded. For a small group of administrators, regular weekly, monthly and annual meetings meant that they developed networks of acquaintances and friends that would have been impossible before the arrival of football. For a smaller group, football provided a vehicle for exploring beyond the boundaries of the coalfield. During the late 1880s several teams from the district competed against leading Scottish teams such as Aidreonians and Motherwell.[57] However, it was Shankhouse Black Watch's entry into the short-lived North Eastern League in 1890 that signalled the entry of coalfield teams into regular competition with teams from Durham and North Yorkshire. This was put on a firmer basis in 1891 with the formation of the NA. Between then and 1914 up to eight teams played in the NA thus providing, to a small number of players, exposure to places beyond the colliery villages. Perhaps even more significant was the entry of Shankhouse British School into the Northumberland and Durham Schools' Cup in 1898. From thenceforth, East Northumberland teams competed in the schools' competitions and after 1906 in competition with teams from as far away as London. Thus for a small group of students the world opened up at an early age. Basically, football served to break down the isolation of the mining villages, bringing increasing numbers of miners into contact with the world outside. In small, subtle ways the horizons of the miners were expanded.

As football served to break down the isolation of the mining villages, it also served to shift the focus of their interests outside the district and allowed them to see themselves as North Easterners rather than as coming from their isolated mining villages. This was reflected, most clearly, in the emergence of Newcastle United as the symbol of the pride of Tyneside. During the 1880s Newcastle West End and Newcastle East End competed against local teams such as Shankhouse, Ashington and Blyth. By the late 1880s both clubs were importing players from Scotland. As a result of the drive to be best both fell into financial problems. This led to their amalgamation in 1892. As Newcastle United they moved to St James' Park and thus began the move towards fame and fortune. After gaining admission to the Football League in 1893, United began to disengage from local competition. As success followed success, they began to be perceived as representative of Tyneside rather than just Newcastle. Promotion to Division One of the Football League brought immediate success. At a pre-season game, in 1898, 22,000 spectators flocked to St James' Park – United had arrived. In the space of less than ten years, United had gone from being a local club to being regarded as the representative of the region. However, it was their amazing record in the first decade of the new century that solidified their hold on the minds and hearts of the inhabitants of Tyneside. Winning the league championship in 1905, 1907 and 1909 and reaching five FA Cup Finals (1905, 1906, 1908, 1910 and 1911), they were the most successful team of the Edwardian era. Miners were amongst the most ardent supporters of United. This is a critically important change that reflected subtle changes in how they perceived themselves and the outside world. In the 1880s, Newcastle were the "enemy". Within 20 years, in certain ways and on certain occasions, they became one with the hated enemy – they became North Easterners. Their perceptions had, in small ways, been changed.

While football provided a focal point for coalfield solidarity, it was also the vehicle for the playing out of inter-village rivalry. Football soon became the focal point of attention that served to exacerbate and intensify local rivalries. By the early 1890s, a number of intense traditional rivalries emerged as the focal point of the football season. In the larger villages, with a more varied occupational base and relatively stable clubs, football became the site of community pride. The colliery managers and local businessmen used football as a visible symbol of community. Such was certainly the case with Ashington FC in the post-1895 era. It was at that juncture that the colliery officials moved to take over the club and develop it as a symbol of community pride. The same was true, at a later date, for Blyth FC and Bedlington United. In all probability, the smaller villages did the same thing. Certainly this was the case in Seaton Delaval by the turn of the century. Football was, in many ways, an ideal vehicle for the cementing of relationships between different segments of the community. It was, in the eyes of the colliery officials, in the final analysis unimportant and ideologically neutral. However, it was one of the few areas of life in which the different segments of society could interact without threatening the basic social relationships which undergirded society.

To the burghers of Newcastle the inhabitants of the mining villages were different. The successive attempts to remove the miners from the Town Moor culminating in the ban of 1880 were all related to the perceptions that the miners were different. So different, uncouth, and unacceptable that the Novocastrians wished to have nothing to do with them. Football added a new dimension and soon after the introduction of football, complaints began to appear in the newspapers and within the boardroom of the NFA. There is no doubt that to the newspaper reporters and the administrators of the NFA the miners, both as players and spectators, engaged in unacceptable behaviour. While the mining clubs accounted for 44 per cent of the membership of the NFA, they accounted for 76 per cent of the suspensions and censures.[58] These complaints provide glimpses into the nature of the mining communities themselves.

One thing never changed during the period to 1914, the nature of the complaints. These are classically illustrated in the history of Shankhouse from its formation in 1884 to its amalgamation with Shankhouse Albion in 1906. Problems emerged in the first season when they travelled to Jesmond for the final of the NFA cup competition. As a result of this game, the city clubs threatened to have police in attendance at the matches – the reason being "That there was some foul language used by supporters of the Shankhouse club."[59] This was the first of what became perennial complaints against the language used both by players and spectators. In the following year the first complaints over rough play were voiced in the press. The fact that these were based on different expectations was demonstrated in 1892 in a game against Blyth. To a Blyth supporter the rough play engaged in by the Shankhouse team culminated in "outright attacks" on the Blyth players which were encouraged by the crowd in calling for them to "go for the man." To the president of the Shankhouse club, it appeared that the complaints were due to the fact that "the 'blacks' are 'Pit Laddies'", "and play a strong game."[60] During the following years there were frequent accusations of "foul play". For example, in 1895 in the semi-final of the NFA Cup against Newcastle United, Shankhouse's centre forward was sent off after "He had previously been cautioned for kicking opponents in a most malicious manner."[61] In the following year, in the semi-final at St James' Park, another player was sent off for "roughness".[62] Later, in 1901, a NA game was delayed for 15 minutes because Shankhouse disputed a goal.[63] In October of the same year, in a game against Newcastle United "A", there was a disturbance over a penalty and the players nearly came to blows.[64] However, the problems did not only come on the playing field; the spectators, too, entered into the fray. Threatening the referee, pitch invasions, foul language, and attacking the referee after the game were common complaints. Finally, on several occasions players and even the whole team marched off the field when they disagreed with the referee's decision.

The example of Shankhouse could be replicated for many teams in the district. The consequences of these complaints, in the Council of the NFA, were closures of grounds for periods varying from one week to two months because of spectator behaviour. In fact, spectator violence reached crisis proportions between 1903 and 1906. Prior to 1903, ground closures were limited to particular grounds, but in 1904 it became a general problem.

> During the season a good many complaints have been made against clubs regarding the conduct of spectators. The practice of commencing hostilities against visiting teams and referees is becoming quite common in some districts of the county.[65]

In fact, the "some districts" were the mining villages of East Northumberland. There was a consistent pattern of player and referee harassment. In 1895, the spectators at Choppington broke onto the playing field and chased the players.[66] In 1909 the police at Seghill were called to protect the players.[67] In 1910, spectators at North Seaton pelted the referee with mud and the Earsdon ground was closed after a series of battles between spectators and opposing players.[68] Player behaviour was dealt with by issuing suspensions varying from one or two games to *sine die*, to strongly worded motions of censure. While similar complaints were made against other teams, the mining clubs were the most obvious and conspicuous offenders. Behaviour on the football field and at the grounds simply reflected the reality of life in the mining villages. Life in the villages and down the mine was characterized by overcrowding in the houses and dangerous conditions at work. Their behaviour on the playing field simply reflected the conditions at work and home. Miners solved many of their problems by fighting. In fact, historically, fighting was a central part of life and the acceptable way of solving problems. Their whole concept of masculinity was rooted in the harsh physical reality of work in the mines. Similarly, their show of independence is not surprising since one of the most common complaints voiced against the miners was their sturdy independence. As for their use of "foul and abusive language," this simply reflected the reality of life. Harsh and dangerous work conditions did not promote the use of "civilized" language. The language used on the field was the language used in daily life. One man's bad language was another's daily speech. In fact, football epitomized some of the most deeply held values of mining life, independence, cooperation, and rugged masculinity. Football brought these basic elements together in a single totality. The role of the individual player paralleled the role of the hewer in the pit. At the same time, cooperation or teamwork was a fundamental aspect of life down the pit. Finally, a man was judged by his physical prowess. On the football field they were simply fulfilling their own sense of being a man.

The history of local football also reveals something about the nature of the mining communities. Football illustrates the very strong sense of community that characterized the mining villages. By the turn of the century, football had, perhaps, replaced the traditional sports as the most important community symbol. Football allowed them to judge themselves against the outside world as well as within the coalfield itself. Not only did football provide a mechanism for judging themselves against others, it also provided a vehicle for supporting community institutions. Football was an ever-present element of community celebrations and was used on many occasions to raise funds for various causes and, in particular, to raise funds for workers injured in the pit. At the same time, their actions on the football field and at the football ground revealed a strong streak of independence, of a group willing to stand up for their rights and willing to accept the consequences of their actions. The miners at play differed little from the miners at work.

The shift from traditional sports to the modern sport of football was indicative of deeper changes that were having an impact upon the mining communities in the later part of the nineteenth century. While the work of the hewers had changed little during the course of the century, the society had been changed, significantly, by the processes of urbanization and industrialization. These, too, had an effect upon the mining villages and football. The development of leagues, regular schedules and district organizations revolutionized their sporting practices. Thus the 1890s represent a watershed in the history of the mining villages. Increasingly time became structured and space became enclosed. The loose time and spatial boundaries of the traditional sports were eroded and replaced by regularly scheduled competitions played in specially created, enclosed grounds. Life in the mining villages was changed; they were no longer the completely isolated villages of the early years of the century. And nowhere is this opening up to the world more clearly demonstrated than in the increasing importance of the schools. The schools brought alien ideas into the classroom and to the youth. No longer did young boys learn solely from their fathers down the mine; now they were also subject to the influences of their teachers.

The villages were, in every sense, dominated by the pit shaft. It was the focal point of village life. The pit whistle signalled whether they would work the next day, signalled a pit disaster, and in general regulated the flow of life in the community. And it was the social relationships flowing from the mine that dominated life within the villages. Football reflected, very clearly, these social relationships. It cut through all levels of society and no group was immune from the football mania. At the same time it reflected the paternalism of the colliery owners and officials on the one hand, and the independence of the miners on the other. The owners and officials provided strong support to the high prestige clubs but were also supporters of various cups and teams. At the same time, miners demonstrated their sturdy independence and as time passed turned, in increasing numbers, to their own institutions to support the game. Throughout, football demonstrated its roots in traditional institutions, in particular the inn. Eventually adopted by the schools and churches it penetrated to every level of mining society. In fact, by the 1920s, it penetrated into the alien world of women with the formation of a number of women's teams.[69] Very simply, football reflected the complex relationships within mining society. While the barriers between different groups were real, they were always porous and communication was always taking place between different groups. It was this complex set of relationships that characterized mining life, not an overly simplified view of two completely separate worlds. The mining villages, truly, were communities whose characters were shaped and contoured by the complex relationships between the different groups. Football, more than any other sport, reflected the complex reality underlying the totality.

One final question remains to be answered, to what degree did the history of football in East Northumberland parallel its history elsewhere? The history of football in East Northumberland was different in two important ways: the timing of events and the groups who initiated the game. East Northumberland lagged behind such places as Sheffield, Birmingham, Blackburn, Manchester, Stirling, Teeside and Nottingham in the development of clubs, leagues and associations.[70] The differences are illustrated, most clearly, in the development of football in the schools. By the early 1890s there were school leagues and associations in many cities.[71] These did not come to East Northumberland until the early 1900s. These developments, in turn, relate to the initial stimulus to the development of the game. In East Northumberland, in the 1880s, the initial stimulus came from within the mining villages, from the miners themselves. This rootedness in mining society was reflected in the strength of tradition – the number of games played in the annual picnics. This was not the case in other places: doctors, vicars and teachers on Teeside;[72] churches/chapels in Sheffield, Nottingham, Liverpool and Birmingham; works in Sheffield and Birmingham.[73] In fact, it was not until the early 1900s that these institutions sponsored teams in the mining villages. For example, in Ashington, after 1900, of 86 teams formed, 10 were associated with churches/chapels, 11 with Miners' Institutes and/or social clubs, 5 with school teams, 4 with temperance societies and 7 with neighbourhoods or public houses. What this suggests is that the stimulus behind the initial development of football in the mining villages was different from that in the larger towns. Football emerged out of mining society and was not introduced by middle-class groups.

Conclusion

The title of this chapter was "Football: a symbol of sport in East Northumberland." To what extent did the history of football parallel that of other sports and what does this suggest about the development of sport? In fact, there were several elements that were common across most sports, to such a degree that it is possible to offer some generalizations about sport in East Northumberland. One of the most important elements was the power of tradition. Tradition was alive and well in 1914, perhaps still more popular than the emerging sports. Quoits, potshare bowling, short distance pigeon racing and even football were still being played in ways their forefathers would have recognized. However, it did appear that they were practised by the "older" men in the villages.

While traditions were alive and well, significant changes took place that heralded the future, a future far removed from the past. Sport, once epitomized by the traditional challenge matches and irregularity, became identified with clubs, leagues and regular schedules. These changes took the games away from the players and into the hands of "administrators". This moved the control of sport into the hands of individuals with some administrative expertise, and it was clubs with these "experts" that gained a degree of stability and permanence and became an ingredient in the new traditions – Backworth Percy, Newcastle United, etc. In some sports they moved under the umbrella of externally controlled organizations and, thus, lost the individuality and independence which characterized sport in the mining villages.

One of the central characteristics of miners' sports was that they were played on public land or spare land in close proximity to inns. This maintained their ability to control their own games. They were gradually forced from public land into enclosed grounds. Sport was played on grounds controlled by landowners, colliery owners and innkeepers. At the same time, the actual control of these groups was moderated by the development of miner-owned facilities. They were, however, gradually removed from their traditional playgrounds.

Finally, by 1914 it could be argued that football was the most popular sport. The shift from traditional sports to football represented the death of tradition in the villages and reveals the most important change that threatened the independence of the villages. Young children learned the ways of manhood from their fathers, first at home and then down the pit. The gradual change of the school leaving age, from 9 to 15, meant that young children became exposed to outside ideas for longer and longer. One of the most important things they learned from the schoolmaster, starting in the mid-1890s, was the game of association football. Thus, when they eventually left school to go down the mine, they had already acquired the game that was to dominate their lives – football.

It is important not to over-emphasize the changes; traditions were still alive and well in 1914. However, it is possible to identify, quite accurately, when the major changes took place. These changes paralleled economic conditions which led to the availability of more "ready money". The early 1870s, the most prosperous era in the nineteenth century, witnessed the creation of an infrastructure of facilities and the first move to enclosed grounds. As the Depression lifted in the mid-1880s, both sport and leisure activities moved into a period of rapid expansion: the variety of programmes in the school halls, the emergence of association football, the inauguration of Children's Day and the opening of the Coop Hall in Seaton Terrace. These changes came to fruition in the mid-1890s in a period of improving economic conditions. It was during this period that most sports acquired the characteristics of modern sport. The world had changed and there was no going back. The mining villages had moved into the modern world.

Conclusion

If there was ever a study that illustrated the validity of Marx's famous dictum "that men make their own history but not necessarily in conditions of their own making," it is this study of sport and leisure in East Northumberland. It has illustrated how miners and their families struggled to create meaningful lives in conditions over which they had little control. What then have we learned about the nature of the villages, the role of sport and leisure in their lives, and how they relate to other working-class enclaves in Britain?

In fact, contemporary views of the villages as comprising marginal housing, poor sanitation and a culture of drunkenness and violence were not without foundation. Living conditions were not conducive to the ideal of family life espoused by middle-class Victorians. In fact, cramped living conditions literally forced men out of the house and into the arms of the innkeepers and inns, havens of warmth, companionship and drink. And there is little doubt that much beer was drunk in the "pubs". At the same time, it is evident that the majority of the population did not drink to excess. As to violence, fighting was a part of life, the accepted way of solving problems. This reality must always be remembered when looking at life in the villages. Thus, it is important to emphasize that the villages were not pleasant places to live and life was a continual struggle. As Jimmy Tabarar commented, they "used to say the good old days but there weren't any."[1] In some ways, the marginal living and working conditions made sport and leisure all the more important since they were the only avenue, apart from religion, to achieving a sense of self and meaning. Their lives were lived within an environment created by the land and colliery owners and on-site colliery officials. Even though there was daily interaction between the different social groups in the villages, the basic reality was clearly understood by all – power lay in the hands of the colliery owners and officials. What is rarely recognized is the fact that generations of miners and their families made the best of their surroundings and created meaningful lives in conditions which, to outside observers, were marginal. The rich and complex leisure activities outlined in the preceding chapters call into question simplistic accounts of the mining villages as dens of vice, drunkenness and violence. What was revealed was an amazingly complex and rich tapestry of activities that covered the whole spectrum from Shaw and Ibsen to melodramas, symphonic music to brass bands, lawn tennis to potshare bowling. In fact, the activities crossed the full spectrum of leisure pursuits from high culture to low. What was truly surprising was that this wide spectrum of activities was available in relatively small mining villages.

What role did sport and leisure activities play in their lives? In the early years they reflected the reality of life: irregular, with times of feast and famine, and based on the availability of "ready" money. More than at any other time in the century the miners controlled their own games. Miners used the inns as the centres for organizing everything. Nearly without exception the contests were between individuals and focused on the ubiquitous challenge matches. More importantly the actual contest was only part of a larger process. Each contest brought groups of people to the inns to make arrangements for the contest, to establish the rules, to negotiate the stakes, to decide on referees, and to draw up articles of agreement. The whole process entailed several meetings. At the designated time the parties met at the inn and then repaired to the location of the contest, invariably on public land. The contest was followed by a return to the "pub" to settle accounts. The central importance of the "pub" cannot be over-emphasized, it was the only institution available for groups to meet to organize activities.

What happened during the course of the century? While traditional games maintained a strong hold until the First World War, there were significant changes, all related to external forces. There were four major changes that channelled sport down new paths and eroded the miners' power to control their own games. First, innkeepers moved from providing a meeting place to becoming major sponsors of events. It was innkeepers who initiated handicaps in the 1850s. By the early 1870s they became the major force behind the commercialization of sport – a process that engulfed nearly all sport. A second change was the movement to force competitions from public land into enclosed grounds. This brought about a significant change – now miners had to pay to watch their sports. Third, as club sport moved to centre stage, the organization moved into the hands of non-players, individuals with some administrative experience. Fourth, as they strove to reach the top they moved within the jurisdiction of regional and national organizations whose base lay outside the colliery villages. These changes eroded their power to control their own sports and brought them into the mainstream of British sport.

Thus, the inhabitants of East Northumberland were inexorably drawn within the wider world of sport and lost control of their own games. In something of a paradox as they were losing control of their games they gained a measure of independence, in fact, greater independence than they had ever had before. The opening of the Coop Hall in Seaton Terrace in 1885 was revolutionary. At last, they had access to space outside the control of the colliery officials. Subsequently, in all villages, buildings were financed, built and owned by miners. This precipitated an explosion of interest in quoits and billiards, the first truly mass sports in East Northumberland. It was within these social clubs that they maintained their traditional practices – an emphasis on intra-club competition and a rejection of leagues. At the same time, they used these facilities to promote the "new" game of association football. Thus, by 1914, their sport was a complex amalgam of the old and the new, of the residual, dominant and emergent.

What was the meaning of sport and leisure to the villagers of East Northumberland? As to their sporting activities, there was no doubt about what it was all about – winning. Until the churches and schools started sponsoring football teams in the 1890s there was no evidence of any concern about character building and the development of other values. In something of a contradiction there was evidence of a deep concern over amateurism. There were amateur bicycle clubs in the 1870s and 1880s, amateur rowing clubs in the early 1900s, amateur brass bands in 1902, amateur boxing, and amateur billiard leagues. However, attempts to be recognized as amateurs foundered on the London-based definition of an amateur. Miners recognized the barriers to their acceptance – the payment of money in any shape or form. How could miners, whose games were played for money and victory, claim to be amateurs? It is clear that they were not interested in definitions of an amateur that included statements like, "Once a professional always a professional" or exclusionary definitions based on class. What they were concerned about was a "level playing field." They wanted to ensure that at the start of any contest conditions were as equal as possible; thus the immense popularity of handicap competitions in potshare bowling, quoits, and foot racing. As far as can be ascertained, "amateurism" did not refer to behaviour on the field; in other words it was the "spirit" of amateurism which was so important to the moguls of amateur sport. When we turn to the playing field, it is evident that miners did not play like "gentlemen". There is no doubt that, to outsiders, miners exhibited unacceptable behaviour on the playing field. At the same time, in something of a contradiction there were many references to miners demonstrating fair play and sportsmanship. This presents us with an enigma. How can the two faces of miners' sport be reconciled? The fact that they did not exhibit the behaviour associated with amateurism does not mean there was not a generally accepted code of conduct. What it does mean is that they were different.

How important were sport and leisure? To large numbers of men sport was one of the most important things in their lives. It was talked about down the pit, at the social clubs and on the street corners. Poets waxed eloquent about sport. It reinforced their view of themselves. Their sports emphasized qualities of strength, endurance, skill and luck, basic elements of mining life. Athletic heroes had great status in the villages and beyond, they carried the pride of the mining communities with them when they competed outside the area. It is difficult to over-emphasize the importance of sport to the villages. In what can charitably be called a harsh living and working environment it was one activity that brought fun, enjoyment and sociability and a never ending topic of discussion. Most importantly it reinforced who they were as men.

Perhaps of greater significance are the similarities and differences between this history and the history of sport and leisure in other working-class enclaves. Much of the history of East Northumberland finds resonance in other parts of the country. Amongst mining communities, throughout Britain, the same games were played: quoits, dog racing, foot racing, pigeon flying, pitch and toss, and football. All these focused on the "pub". Club sport came to all working-class areas and with it affiliation with district and national organizations. Similarly, sport was forced from public land into the confines of enclosed grounds owned and operated by different groups. As in East Northumberland, sport had to be financially viable. This was not easy to achieve in working-class districts and, thus, the massive instability that characterized sport. With respect to non-sporting leisure activities, all the activities popular in East Northumberland were to be found in the working-class enclaves of the cities, towns and villages. However, this work supports Williams' contention "That there was no over riding sense of working class identity to which all working class people subscribe."[2] In fact, the differences between working-class groups were, in many respects, greater than the similarities. The similarities tended to be on the surface. We need to explore beneath the surface if we are truly to understand the history of sport and leisure in Britain.

On the surface, the individual differences appear to be of small concern, but when added together they raise some interesting questions about the history of sport and about British sport history. One of the most obvious differences was the timing of events. The mining villages lagged far behind other areas in the development of sport. Second, while the institutions that sponsored sport were drawn from the same groups, the mix of institutions sponsoring sports in different locations differed greatly. Third, facilities available for sport varied tremendously – few working-class groups had access to a Town Moor or beaches. Fourth, the particular form of particular games varied from district to district, for example the rugby and association football zones in Yorkshire. Finally, while amateurism was present in nearly all working-class areas, the approach taken differed from area to area. When put together these differences lead to the conclusion that the differences were, in many respects, greater than the similarities.

It is apparent that in some instances the differences were greater than the similarities. Different sports had different patterns of development, sponsorship and participants. Even within sports there were differences in their patterns of development. In other words, development differed from locality to locality, from region to region. This supports Huggins' statement about the "Neglected importance of regional and local sporting identity."[3] Much of the work in the 1980s attempted to look at sport from a "national" perspective. In so doing great attention was paid to amateurism, professionalism and commercialization. This served to obscure the rich and varied regional and local histories. In fact, several historians have focused on local or regional themes: Huggins, Hill, Williams, Speak, etc. At the same time few historians, with the notable exception of N. Tranter, have examined the totality of sporting experiences in a particular geographical location. What this study of a small area defined by coal mining has done is to point to the immense complexity of sport and leisure in the area. The differences were so great that villages only a mile or so away from each other had significantly different sporting histories – some had none. If we are truly to understand the history of sport in Britain we must explore regional and local sport history. Very simply we need in-depth examinations of local and regional sport history before we can synthesize them into a truly national picture. A start has been made with some work on the textile districts, the North East of England, and Central Scotland. Now we need to go further.

Plate 7.1 Barrington Football Club, 1921.
Reprinted with permission from Martin, E. (1997) *Images of England: Bedlingtonshire*, Stroud: Tempus Publishing

One group has been notable by its absence from this history – women. This was deliberate because *if* they had any leisure it could not be examined within the context of the definition used here – free time. If young women had leisure it was "interwoven with their day-to-day routines and social and familial responsibilities." Wives and mothers, on the other hand, carved out "spheres of pleasure in lives dominated by work and worry." Additionally, mining society was very much a male society. Thus, the examination of females would require a totally different approach. Unfortunately this lies outside the scope of this monograph. However, changes did take place in the mining village. First, within the context of church and chapel, women did get involved. With the rise of commercial entertainment they were enthusiastic supporters and often performers. Their lives certainly changed in the early 1900s with the explosion of interest in dancing. What this suggests is that changes were taking place. The history of the women in the colliery villages awaits a researcher.

Notes

Series editors' foreword

1 Alan Metcalfe, *Leisure and Recreation in a Victorian Mining Community: The Social Economy of Leisure in North-East England, 1820–1914* (London: Routledge, 2005), p. 197.
2 William J. Bouwsma, "From History of Ideas to History of Meaning," in Theodore K. Rabb and Robert I. Rotberg (eds), *The New History and Beyond* (Princeton: Princeton University Press, 1982), p. 288.
3 Arthur Bryant, *English Saga 1840–1940* (London: The Reprint Society, 1942), p. 231.
4 C. Aspin, *Lancashire: The First Industrial Society* (Helmshore: Helmshore Local History Society, 1969), p. 180.
5 Metcalfe, *Leisure and Recreation*, p. 160.
6 Bernice Martin, *A Sociology of Contemporary Cultural Change* (Oxford: Blackwell, 1981), p.2.
7 Metcalfe, *Leisure and Recreation*, p. 163.
8 Martin J. Weiner, *English Culture and the Decline of the Industrial Spirit 1850–1980* (Cambridge: Cambridge University Press, 1984), p. 7.

Introduction

1 J. Burnett in Introduction to *John Wilson, Memories of a Labour Leader* (Sussex, 1980) p. viii.
2 *Newcastle Daily Chronicle* (*NDC*), 17 December 1909.
3 N. McCord, *North East England. The Region's Development, 1760–1960* (London, 1979), p.191.
4 *Newcastle Weekly Chronicle* (*NWC*), 26 January 1895.
5 *NDC*, 5 August 1907.
6 For the most up-to-date discussion of women's leisure see Catriona M. Parratt, "Little Means or Time: Working-Class Women and Leisure in Late Victorian and Edwardian England," *International Journal of the History of Sport* (*IJHS*) 15.2 (August, 1998), pp.22–53.
7 H.E. Meller, *Leisure and The Changing City, 1870–1914* (London, 1976); J. Lowerson and J. Myerscough, *Time to Spare in Victorian England* (Brighton, 1977); P. Bailey, *Leisure and Class in Victorian England: Rational Recreation and the Contest for Control, 1830–1885* (London, 1978); J. Walvin, *Beside the Seaside: A Social History of the Popular Seaside Holiday* (London, 1978); *idem, Leisure and Society, 1830–1850* (London, 1978); H. Cunningham, *Leisure in the Industrial Revolution, c.1780–c.1880* (London, 1980); *idem*, "Leisure" in J. Benson (ed.), *The Working Class in England, 1875–1914* (London, 1985), Chap.5, pp.133–164; *idem, Leisure and Culture* (London, 1990).
8 J.A. Mangan, *Athleticism in the Victorian and Edwardian Public School* (Cambridge, 1981).
9 D. Birley, *Sport and the Making of Britain* (Manchester, 1993); *idem, Land of Sport and Glory. Sport and British Society, 1887–1910* (Manchester, 1995).

10 J. Lowerson, *Sport and the English Middle Classes, 1870–1914* (Manchester, 1993).
11 T. Mason, *Association Football and English Society, 1863–1914* (Brighton, 1980); W. Vamplew, *The Turf: A Social and Economic History of Horse Racing* (London, 1976), M. Huggins, *Flat Racing and British Society, 1790–1914: A Social and Economic History* (London, 2000); E. Halladay, "Of Pride and Prejudice: The Amateur Question in English Nineteenth Century Rowing," *IJHS* 4.1 (May, 1987), pp.39–54; N. Wigglesworth, *Victorian and Edwardian Boating* (London, 1987); *idem, A Social History of English Rowing* (London, 1992); D. Birley, *A Social History of English Cricket* (London, 1999).
12 T. Mason, *Sport in Britain* (London, 1988), R. Holt, *Sport and the British: A Modern History* (Oxford, 1989); N. Wigglesworth, *The Evolution of British Sport* (London, 1996).
13 R. Holt (ed.), *Sport and the Working Class in Modern Britain* (Manchester, 1990), p.1.
14 *Ibid.*
15 J. Hill and J. Williams (eds), *Sport and Identity in the North of England* (Keele, 1996).
16 There is only one scholar who has attempted a similar project, Neil Tranter and his work on 18 outdoor sports in Central Scotland. His outstanding research is different from this on several counts. First, it only deals with outdoor sports. Second, the area was different from East Northumberland in that it had very different economic, social and demographic conditions. His corpus of material is to be found in the following: N.L. Tranter, "Popular Sports and the Industrial Revolution in Scotland: The Evidence of Statistical Accounts," *IJHS* 4.1 (May 1978), pp.21–38; *idem*, "The Social and Occupational Structure of Organized Sport in Central Scotland During the Nineteenth Century," *IJHS* 4.3 (December, 1987), pp.301–314; *idem*, "Organized Sport and the Middle-class Woman in Nineteenth Century Scotland," *IJHS* 6.1 (May 1989); *idem*, "The Chronology of Organized Sport in Nineteenth Century Scotland: A Regional Study I – Patterns," *IJHS* 7.2 (September 1990), pp.188–203., *idem*, "The Chronology of Organized Sport in Nineteenth Century Scotland: A Regional Study II – Causes," *IJHS* 7.3 (December, 1990); *idem*, "Organized Sport and the Working Classes of Central Scotland, 1820–1900: The Neglected Sport of Quoiting" in R. Holt (ed.), *Sport and The Working Class in Modern Britain* (Manchester, 1990), Chap.3, pp.45–66.
17 A loose translation of K. Marx, "The Eighteenth Brumaire of Louis Bonaparte" in Lewis S. Feuer (ed.), Marx and Engels, *Basic Writings on Politics and Philosophy* (New York, 1959), p.320.
18 For the most complete analysis of the characteristics of modern sport see A. Guttmann, *From Ritual to Record: The Nature of Modern Sport* (New York, 1978).

1 The social context: East Northumberland, 1820–1914

1 This brief history of the Ridleys was compiled from a variety of sources: R.E. Gerrard, *Northern Notables* (Newcastle, 1987); material gleaned from the *Newcastle Daily Chronicle* (*NDC*), *Newcastle Weekly Chronicle* (*NWC*), *Blyth News and Weekly Transcript* (*BNWT*) and *Morpeth Herald* (*MH*); documents in the Ridley Papers at the Northumberland Record Office (NRO); and Minute Books of Blyth Local Board, Blyth Urban District Council, Cramlington Urban District Council, Cowpen Urban District Council and Seaton Delaval Parish Council.
2 C.E. Baldwin, *The History and Development of the Port of Blyth* (Newcastle, 1929), p.62.
3 Information on the Samples and the Duke of Portland's involvement in the area is drawn from the Sample Papers, NRO.
4 Data is derived from *Kelly's Directories of Northumberland and Durham*, 1858, 1873, 1887, 1894, 1902, 1910, 1914.
5 *Kelly's*, 1902.
6 *Kelly's*, 1902, 1910, 1914.
7 Sample Papers.
8 *NWC*, 17 May 1873.
9 *NWC*, 23 August 1873.

10 R. Redmayne, *Men, Miners, and Memories* (London, 1942), p.23.

11 Memoir of John George Weeks. Read before the North of England Institute of Mining and Mechanical Engineering by R.J. Weeks, 9 February 1918.

12 Minutes of Cramlington Local Board, 1865 to 1890.

13 The consequences of this Act and the control of the land and mine owners is discussed in J.M. Taylor, *England's Border Country: A History of Northumberland County Council, 1889–1989* (Morpeth, 1989), pp.6–47.

14 For the importance of this Act see Harold J. Laski, W. Ivor Jennings and William A. Robson (eds), *A Century of Municipal Progress. The Last Hundred Years* (London, 1976), pp.49–52.

15 See Minutes, Seaton Delaval Parish Council, 1895–1912: 18 April 1906, 13 June 1906, 3 July 1907, 4 September 1907.

14 There are a limited number of sources that are essential reading for anyone interested in coal-miners. Foundational reading for British coal-mining and the conditions of living and working are J. Benson, *British Coalminers in the Nineteenth Century: A Social History* (New York, 1980); M. Pollard, *The Hardest Work Under Heaven. The Life and Death of the British Coal Miner* (London, 1984). For the North East coalfield a basic introduction is N. McCord, *North-East England: An Economic and Social History* (London, 1979). For the North East Coalfield in particular: R. Colls, *The Collier's Rant: Song and Culture in the Industrial Village* (London, 1977); *The Pitmen of the Northern Coalfield: Work, Culture and Protest, 1790–1850* (Manchester, 1987); R. Moore, *Pit-men, Preachers and Politics. The Effect of Methodism in a Durham Mining Community* (Cambridge, 1974); W. Williamson, *Class, Culture and Community. A Biographical Study of Social Change in Mining* (London, 1982); E. Welbourne, *The Miners' Union of Northumberland and Durham* (Cambridge, 1923); R. Fynes, *The Miners of Northumberland and Durham* (Newcastle, 1986; reprint of book published in 1873).

17 The data for this analysis of Seaton Delaval is taken from the Census, 1841–91.

18 1891 Census, Seaton Delaval.

19 The data was taken from the Census, 1841–91 and the *Newcastle Daily Chronicle.*

20 There is no single source that brings together the material on living conditions in working-class areas. The data, for this analysis, was drawn from the following sources: A. Briggs, *Victorian Cities* (London, 1971); D. Cannadine and D. Reader (eds), *Exploring the Urban Past: Essays in Urban History by H.J. Dyos* (Cambridge, 1982); H.J. Dyos, *Victorian Suburb: A Study of the Growth of Camberwell* (Leicester, 1966); T.C. Barker and J.R. Harris, *A Merseyside Town in the Industrial Revolution, St. Helens, 1750–1900* (London, 1959); G.S. Jones, *Outcast London* (London, 1984); R.A. Church, *Economic and Social Change in a Midland Town. Victorian Nottingham, 1815–1900* (London, 1966); M. Harloe, *Swindon: A Town in Transition* (London, 1975); G.E. Cherry, *Birmingham: A Study in Geography, History and Planning* (Chichester, 1994); R. Dennis, *English Industrial Cities of the Nineteenth Century. A Social Geography* (Cambridge, 1984); J. Benson, *The Working Class in Britain, 1850–1939* (Cambridge, 1984); S. Meacham, *A Life Apart. The English Working Class, 1890–1914* (London, 1977); N. McCord, *North East England. The Region's Development, 1760–1960* (London, 1979).

21 Data for South Blyth, Cowpen and Newsham from *Kelly's Directory of Northumberland and Durham*, 1858.

22 S.B. Martin, *Netherton (Nederton)* (Bedlington, n.d.), p.12.

23 *Ibid.*, p.15.

24 S.B. Martin, *Sleekburn (The Station)* (Bedlington, n.d.) p.8.

25 Formation of Allotment Committees: Minute Book, Earsdon Urban District Council, August 1907–July On the 10, 21 February 1908; Minute Book, Bedlingtonshire Urban District Council, 1910–11. First reported Standing Committee on Allotments, 1910; Minute Book, Cramlington Urban District Council, 1911–13, 25 April 1911; Minute Book, Blyth Urban District Council, December 1911–June 1913, 9 May 1912.

26 *NWC*, July 1873.

27 *NWC*, July 1873.

28 1891 Census, Seaton Delaval, West Holywell, New Hartley, East Holywell, Seghill, Bates Cottages, Old Hartley.
29 Data from Seaton Delaval Census, 1841–91.
30 In fact, the Census of 1891 gives the population of Seaton Delaval as 4,057. This comprised two separate villages, New Hartley and Seaton Delaval. This analysis is only concerned with Seaton Delaval.
31 Seaton Delaval Census, 1891.
32 Census, 1841–91.
33 Census, 1891.
34 Taped interview with Billy Wilson, Cramlington on 6 July 1976. This is one of 21 tapes of miners and miners' wives that were recorded in 1976. Eight were born in the 1880s, six in the 1890s, four in the 1900s and three had no date. These tapes are in the Northumberland Record Office.
35 Taped interview, Jack Dorgan, Ashington.
36 Taped interview, James Oswell, Ashington.
37 Data for this analysis of wages is drawn from B.R. Mitchell, *Economic Development of the British Coal Industry, 1800–1914* (Cambridge, 1984). Also from reports in the *NDC, MH, BNWT*.
38 *NWC*, 17 May 1873.
39 *NWC*, 17 May 1873, 26 April 1873, 19 July 1873, 20 September 1873.
40 S. Martin, *Bomarsund* (Bedlington, 1987), p.21.
41 Status was very important in the mining villages. This analysis is based on those who assumed leadership roles in various leisure activities.
42 *NDC*, 3 June 1898.
43 The 1870s were a critical era in terms of the necessary conditions for leisure. Several independent events coalesced to provide these conditions.
44 *NWC*, 3 June 1898.
45 Mitchell, *op.cit.*, p.159.
46 *Ibid.*, p.194–195.
47 Comparative data from J. Benson, *The Working Class in Britain, 1850–1939* (London, 1994), pp.39–71; S. Meacham, *A Life Apart: The English Working Class, 1890–1914* (London, 1977), pp.70–71, 214–216.
48 For a detailed analysis of the power relations underlying the creation of facilities see A. Metcalfe, "Sport and Space: A Case Study of the Growth of Recreational Facilities in East Northumberland, 1850–1914," *IJHS* 7.3 (December, 1990), pp.348–364.
49 Data for recreational facilities was derived from *NDC*, 1858–1914; *NWC*, 1850–1914; *MH*, 1858–1900, 1911–14; *BNWT*, 1870–1914; *Kelly's Directories of Northumberland and Durham*, 1858, 1873, 1887, 1894, 1902, 1910, 1914; Minute Books, Bedlingtonshire Urban District Council (UDC) 1893–97, 1897–1900, 1910–11, 1913–14; Minute Books, Cowpen Local Board, 1875–80, 1889–91; Minute Books, Cowpen UDC, 1895–96, 1897–98, 1898–1901, 1901–03, 1903–05, 1905–07; Minute Book, Cramlington UDC, 1903–08, 1908–11, 1911–13; Minute Book, Earsdon UDC, 1903–07, 1907–10, 1910–13; Minute Book, Seaton Delaval Parish Council, 1895–1912; Minute Book, Seaton Delaval UDC, 1912–14; Minute Book, Seghill UDC, 1905–14; Minute Book, South Blyth UDC, 1896–99, 1899–1902, 1902–05, 1905–07; Minute Book, Blyth UDC, 1907–09, 1909–10, 1911–13, 1913–15; Minute Book, Newbiggin-by-the-Sea, 1894–98, 1898–1902; Minute Book, Newbiggin UDC, 1902–05, 1906–09, 1909–13.
50 The period 1868–75 was critical to the development of leisure activities. See A. Metcalfe, "Organised Sport in the Mining Communities of South Northumberland, 1800–1889," *Victorian Studies* 25.4 (Summer, 1982), pp.469–495.
51 *NDC*, 11 April 1904.
52 *NDC*, 27 November 1911.
53 *NDC*, 6 February 1914. Extract from "Club and Institute Journal."
54 S. Martin, *Barrington (Bedlington)*, pp.29, 33–39.

55 Memorandum and Articles of Association of the Ashington and District Working Men's Social Club Ltd., 18 January 1902.
56 *NDC*, 19 March 1909.
57 The development of the system of social clubs was outlined in a series of articles entitled "Club and Pub" in the *NDC*, 8, 9, 10 April 1902.
58 *NDC*, 16 January 1911.
59 *BNWT*, 8 February 1910.
60 M. Kirkup, *The Biggest Mining Village in the World: A Social History of Ashington* (Morpeth, 1993), p.25.
61 The database on inns was drawn from *NDC, NWC, BNWT, MH*. And *Kelly's Directories of Northumberland and Durham*.
62 See Metcalfe, "Organised Sport in the Mining Communities of South Northumberland, 1800–1889," p.483.
63 A. Metcalfe, "Sport and Space: A Case Study of the Growth of Recreational Facilities in East Northumberland, 1850–1914," *IJHS* 7.3 (December, 1990), p.354.
64 *Ibid.*, p.348.
65 *Newcastle Journal*, 6 January 1844.
66 *NWC*, 8 August 1851, 2 September 1851, 3 October 1851.
67 *MH*, 24 October 1891. A case at Moot Hall Police Court, Newcastle over 12 men racing from North Seaton to Morpeth and back.
68 Metcalfe, "Sport and Space," p.350. In fact, efforts were made to ban bowling starting in 1822. A series of prosecutions in the 1850s effectively removed the game from the roads. From 4 February 1874, efforts were made by Newcastle Council to ban bowling on the moor. These efforts culminated in the ban of 1880. By 1881 the miners were back on the moor. Three further attempts were made prior to 1914 – all failed. This is a classic illustration of the limitations of power.
69 *MH*, 17 May 1873.
70 *MH*, 30 May 1874.
71 Dowson was found guilty of allowing betting at his trial at Bedlington Petty Sessions, 29 July 1876. The owners of Killingworth New Grounds were found guilty of allowing betting at their trial at Moot Hall Police Court, Newcastle, 24 June 1876. This was the effective end of the commercial grounds.
72 Opened Saturday 6 August 1881 (*NDC*, 8 August 1881).
73 S. Martin, *Sleekburn (The Station) with Bank Top, East Sleekburn and Red Row* (Bedlington, n.d.), pp.1923.
74 This analysis of the growth of public facilities is based on the Minute Books outlined in note 49.
75 Minute Book, Seaton Delaval Parish Council, 5 June 1901.
76 *NDC*, 23 June 1905.
77 This discussion is based on the books outlined in note 16.
78 1891 Census, Seaton Delaval.
79 J. Benson, *British Coalminers in the Nineteenth Century* (London, 1980), pp.91–92.
80 For example, the impact of the Public Health Act of 1872 which led to the establishment of sanitary authorities across the country. See G.E. Cherry, *Birmingham. A Study in Georgaphy, History and Planning* (Chichester, 1994).
81 In other words, change originated in the areas that first felt the brunt of the industrial revolution.
82 G. Stedman Jones, "Working-class Culture and Working Class Politics in London, 1870–1900: Notes on the Remaking of a Working Class," *Journal of Social History* (1974), p.479.

2 Non-sporting leisure during the nineteenth century

1 Jonathan Rose, *The Intellectual Life of the British Working Classes* (London, 2001), p.240.

2 This chapter is written within the context of the writings on leisure contained in the following: P. Bailey, *Leisure and Class in Victorian England: Rational Recreation and the Contest for Control, 1830–1885* (London, 1978); J. Clarke and C. Critcher, *The Devil Makes Work. Leisure in Capitalist Britain* (Chicago, 1985); H. Cunningham, *Leisure and the Industrial Revolution, c.1780–c.1880* (London, 1980); H. Cunningham, *Leisure and Culture* (London, 1990); J. Lowerson and J. Myerscough, *Time to Spare in Victorian England* (Sussex, 1977); R.W. Malcolmson, *Popular Recreations in English Society, 1700–1850* (Cambridge, 1973); H.E. Meller, *Leisure and the Changing City, 1870–1914* (London, 1976); M. Smith, S. Parker and C.S. Smith (eds), *Leisure and Society in Britain* (London, 1973); J.K. Walton, *The English Seaside Resort. A Social History, 1750–1914* (Leicester, 1983); J. Walvin, *Leisure and Society, 1830–1950* (London, 1978).

3 Self-improvement has received the greatest attention. This is exemplified in P. Bailey, *Leisure and Class*.

4 The influence of Methodism in the North East coalfield has been well documented in R. Moore, *Pitman, Preachers and Politics: The Effect of Methodism in a Durham Mining Community* (Cambridge, 1974).

5 Evidence to J.R. Leifchild, Commission on Children's Employment in the Mines, 1842, p.588.

6 Taken from evidence given by Charles Carr Esq., Proprietor of Seghill Colliery to the Mining Commissioners, "Report of Mining Population in various parts of the Mining Districts," 1846, p.27.

7 Report of the Mining Commissioner, 1850, p.44.

8 Carr's submission to Mining Commissioners, 1846.

9 *NDC*, 13 March 1880.

10 *NDC*, 4 October 1884.

11 This reveals, as clearly as any example, the rejection of owner/management initiatives. It is critical to recognize the importance of this point. The majority of miners rejected anything that was promoted by management.

12 *BNWT*, 31 October 1891.

13 *BNWT*, 13 February 1892.

14 *BNWT*, 7 January 1896.

15 *BNWT*, 28 February 1896.

16 The autodidactic tradition is covered, in depth, in Jonathan Rose, *The Intellectual Life of the British Working Classes*.

17 Laski-Holmes letters. Two letters by Harold Laski written to Justice Holmes in September 1924 and September 1925, after his visits to Ashington.

18 This is a world that, for the most part, lies hidden from view, and yet to ignore it would provide a particularly biased and incomplete view of the lives and interests of miners.

19 *BNWT*, 28 April 1888.

20 *MH*, 6 September 1890.

21 *NDC*, 29 May 1914.

22 BNWT, 24 January 1896; 23 June 1896; 3 July 1896; 18 September 1896; 6 November 1896.

23 *NDC*, 21 January 1904.

24 *NDC*, 10 October, 12 December 1910.

25 This data is taken from a review of 21 tapes of miners/miners' wives born between 1880 and 1910. The interviews were done between 1971 and 1976 by two history graduates from Newcastle University. They are a veritable fount of information. One of the only times one can actually *hear* the miners speak. They are in the Northumberland County Record Office.

26 *NDC*, 24 March 1871.

27 *NDC*, 25 November 1882, 25 September 1886, 30 June 1888.
28 *MH*, 11 July 1891.
29 *NDC*, 8 May 1896.
30 *BNWT*, 30 August 1884.
31 *BNWT*, 30 June 1888.
32 *NDC*, 30 August 1890.
33 *NDC*, 30 July 1894.
34 The reader will have noticed the nearly complete dependency upon newspapers. This reveals the very real problems of getting information on the miners' lives – they kept very few records.
35 *NDC*, 20 June 1874.
36 *NDC*, 17 September 1881.
37 *BNWT*, 27 February 1886.
38 *BNWT*, 19 March 1892.
39 *MH*, 26 March 1892.
40 Data from Minute Book, Seaton Delaval Parish Council, 1895–1908.
41 *MH*, 6 November 1886.
42 *NDC*, 26 May 1896.
43 *BNWT*, 30 July 1892.
44 Again this is the tip of the iceberg. It suggests ongoing involvement in initiatives aimed at improving conditions in the villages. The miners and their wives were pro-active.
45 The following account is taken from material gathered from the *NDC, MH, BNWT*, 1876–1913.
46 *NDC*, 19, 26 September, 3, 10, 17 October 1913.
47 The account is drawn from the Minute Book of Bedlington Brass and Reed Band, 1890–99.
48 Ashington Collieries Deputations Books, 1893–94, 19 January, 9 February, 9 March, 4 April, 13 June 1893.
49 Taped interview, Jack Dorgan.
50 Evidence suggests they always danced; in fact, dancing was one of their favourite amusements. See G. Halliwell, "Fifty Years On," *NWC*, 19 January 1895. "One of the favourite amusements of pitmen, forty or fifty years ago, was dancing at the public houses. These, however, only came round occasionally." Thus, the new quadrille assemblies were simply the formal organization of practices deeply rooted in mining society.
51 *MH*, 17 February, 24 February, 1894.
52 Taped interview, James Oswell.
53 Taped interview, Billy Wilson.
54 M. Kirkup, *A Social History of Ashington* (Morpeth, 1993), p.212
55 *NDC*, 11 February 1882.
56 *MH*, 27 February 1886.
57 *MH*, 11 February 1888.
58 *School Correspondents Record Book for Schools in the Seaton Delaval Area, 1903–*, New Hartley Colliery School, 20 November 1903–9 December 1910.
59 The villages were Seaton Delaval, New Hartley, East Holywell, Holywell, Bates Cottages, Seghill and Cramlington. This supports the contention that it is inappropriate to consider Seaton Delaval in isolation.
60 *MH*, 9 January 1886.
61 1891 Census, Seghill.
62 *MH*, 9 December 1910.
63 *BNWT*, 5 January 1884.
64 *BNWT*, 24 March 1896.
65 *NDC*, 11 February 1888.
66 *BNWT*, 13 December 1884, 4 October 1890.

67 *BNWT*, 20 October 1894.
68 *BNWT*, 14 August 1886.
69 *MH*, 19 January 1912.
70 *Seghill Parish Magazine*, July, 1902.
71 *Seaton Delaval and New Hartley Conventional Magazine*, 1913–1915.
72 Drawn from accounts in the *NDC*, *BNWT* and *MH*.
73 *MH*, 23 May 1891.
74 For an analysis of the growth of the railways and their impact upon life in the villages see J.A. Wells, *The Blyth and Tyne, Part I* (Morpeth, 1989) and *The Blyth and Tyne Branch, 1874–1989* (Morpeth, 1990).
75 *BNWT*, 2 June 1896, 9, 16 August 1890.
76 None of these events were reported in the newspapers until two years later, as they became recognized as important events in village life.
77 *MH*, 20 June 1913.
78 *BNWT*, 24 April 1886.
79 *NDC*, 20 December 1912.
80 I use inns and pubs synonomously. From visiting over 50 of the inns it is obvious that they were all very similar. They were referred to by contemparies as inns or pubs.
81 The histories of the Astley Arms and Hastings Arms have been compiled from newspaper research.
82 Letter from John Robinson on Jubilee of Seaton Delaval Primitive Methodist Chapel. Published in *NWC*, 30 November 1895.
83 *MH*, 3 December 1892.
84 All flower shows were organized at an inn.
85 Examples include meetings with reference to the Cramlington Strike, 1865; Northumberland Miners Mutual Conference Association, 1870; working conditions, 1877. It was also the headquarters for the soup kitchen in 1878.
86 *MH*, 1898.
87 *MH*, 26 October 1895.
88 See J. Benson, *British Coalminers in the Nineteenth Century: A Social History* (London, 1980), p.81, pp.152–154.
89 Rose, *The Intellectual Life of the British Working Classes*.
90 *Ibid*., p.131.
91 Eileen Yeo, "Culture and Constraint in Working Class Movements, 1830–1855," in E. Yeo and S. Yeo (eds), *Popular Culture and Class Conflict, 1590–1914* (Sussex, 1988), p.161; A. Briggs, *Victorian Cities* (London, 1968), p.187.
92 See Bailey, *Leisure and Class*, Chap.5, "Rational Recreation in Operation: The Working Men's Club Movement"; S. Meacham, *A Life Apart: The English Working Class, 1890–1914* (London, 1977), p.121.
93 Rose, *The Intellectual Life of the British Working Classes*, p.237.
94 G.S. Jones, *Outcast London* (London, 1984), p.173.
95 Gareth Stedman Jones, "Working Class Culture and Working Class Politics in London, 1870–1900: Notes on the Remaking of a Working Class," *Journal of Social History* 7 (1974), p.479.

3 The strength of tradition

1 What makes this chapter so difficult to write is that very few sport historians have addressed these sports. Only the upper-class-dominated sport of fox hunting has received attention. Thus, the comparative data is rather thin. There are some exceptions that provide tantalizing insights: R.W. Malcolmson, *Popular Recreations in English Society, 1700–1850* (Cambridge, 1973); D.A. Reid, "Beasts and Brutes: Popular Blood Sports c.1780–1860," in R. Holt (ed.) *Sport and the Working Class in Modern Britain* (Manchester, 1990), Chap.1, pp.12–28. Most of these end at 1850.

2 The whole discussion on the transition from traditional to modern sport is based on the influential work by A. Guttmann, *From Ritual to Record: The Nature of Modern Sports* (New York, 1978).
3 "Cock-Fighting in Newcastle," *North Country Lore and Legend* (March, 1887).
4 *NWC*, 9 February 1895, letter from Geo. R. Keel saying cock fighting was alive and well in Northumberland; *MH*, 2 April 1893; *MH*, 25 June 1902. For cock fighting elsewhere see Reid, "Beasts and Brutes," pp.18–23.
5 *BWNT*, 28 May 1892.
6 D.C. Ilzkowitz, *Peculiar Privilege: A Social History of English Foxhunting, 1753–1885* (Sussex, 1977).
7 Otter hunting was practised on the River Wansbeck between April and September. It was only practised by the Ridleys and the local elite. See *Local Records*, 29 May 1911; *NDC*, 15 July 1902.
8 Ridley Papers, Diary of Sir M.W. Ridley, Hounds, 1830–33.
9 *MH*, 21 October 1871.
10 *MH*, 11 February 1871.
11 *MH*, March 1872.
12 Cramlington Coursing Club Meet, 11–12 October 1912 (*NDC*, 24 September 1912); North Gosforth Coursing, Workmen Stakes (*NDC*, 17 February 1892).
13 *NDC*, 5 April 1892.
14 *NDC*, 16 November 1891.
15 Database from 35 issues of the *NDC* between 1 January 1876 and 16 November 1876.
16 *NDC*, 1 January 1876.
17 *NDC*, 16 June 1876.
18 *Newcastle Journal*, 19 January 1880.
19 Data from *NDC*, September 1912 to April 1913.
20 This data is taken from 17 tapes of interviews with old miners in the period 1971 to 1978.
21 *NWC*, 13 April 1895.
22 Ridley Papers, Game Books, Blagdon, 1864–75.
23 Ridley Papers, Game Books, Blagdon, 1905–14.
24 Sample Papers, Game Book, 1913–14.
25 Database created from *NDC*, *MH*, and *BNWT*.
26 Data from the 17 interview tapes.
27 *NDC*, 27 January 1896.
28 Derived from *NDC*, 1900–10.
29 For pigeon shooting see M.A. Kellett, "The Power of Princely Patronage: Pigeon Shooting in Victorian England," *IJHS* 11.1 (April, 1994), pp.63–85; N. Tranter, *The Chronology of Organized Sport in Nineteenth-Century Scotland: A Regional Study I – Patterns*," *IJHS* 7.2 (September, 1990), pp.190, 191–192, 194, 196, 198.
30 Tranter, *The Chronology of Organized Sport*, p.192.
31 Taped interview with Ned Doyle.
32 Taped interview with James Wilson.
33 Taped interview with Jimmy Tabarar.
34 Taped interview with Huey Smith.
35 For a view of handball in the Durham coalfield see J. Melling, "Sport, Spectacle and Class: Handball in the North East," *IJHS* 7.3 (December, 1990), pp.443–447.
36 *MH*, 16 October 1886.
37 Data from *MH* and *NDC*.
38 This is supported by the fact that all the miners (5 of 17) for whom handball was important came from two areas – Seaton Delaval and Ashington.
39 As late as 1912, the Durham coalfield boasted nine ball alleys that regularly attracted large numbers of spectators (*NDC*, 1912).
40 For a more complete view of potshare bowling see A. Metcalfe "Potshare Bowling in

the Mining Communities of East Northumberland, 1800–1914," in R. Holt (ed.), *Sport and the Working Class in Modern Britain* (Worcester, 1990), Chap.2, pp.12–28.

41 *NDC*, 10 February 1902.

42 This list was drawn up from references in the *NDC*, 1900–05. They were recognized figures in the world of potshare bowling.

43 In fact, according to the *NDC*, handicaps were initiated by Mr Harry Wardle, a Newcastle publican, sometime in the 1840s or 1850s. *NDC*, 30 September 1867.

44 *MH*, 26 November 1898.

45 From *NDC* and *MH*, 1884.

46 From *NDC* and *MH*, 1893.

47 References to Will Summers were spasmodic yet it is apparent that he was recognized as *the* champion. This says more about the lack of media interest. It is difficult to understand why the papers reported potshare bowling extensively but virtually ignored quoits which was, perhaps, the more popular participant sport.

48 *MH*, June, July, August, 1893.

49 *NDC*, 1 September 1908.

50 *NDC*, 18 September 1901.

51 Taped interview Jimmy Tabarar.

52 *NDC*, 15 June 1908.

53 This account was drawn from the *NDC*, 31 May 1898 to 22 August 1898.

54 *NDC*, 8 June 1909.

55 Poem "Pit Life in 1893–94," in Matthew Tate, *Poems, Songs and Ballads* (Blyth, 1898), p.30.

56 M. Tate, "Pit Life in 1893–94," p.28–30; Joseph Skipsey, "The Collier Lad," in J. Skipsey, *Carols from the Coalfield and Other Songs and Ballads* (London, 1886), p.66.

57 Ten of the 17 mentioned them – seven potshare bowling, seven quoits, and five handball. All mentioned at least two of the games.

58 Data from *NDC*, 1886.

59 Data from *NDC, MH, BNWT*.

60 *MH*, 18 December 1897.

61 The evidence suggests that these traditional games were immensely popular in working-class districts across Britain, yet they are notable by their absence from works that claim to be giving an overview of British sport. Except for the notable exception of N. Tranter, no scholars have addressed these sports. Hegemony, a much abused word, is alive and well in academe.

62 Except for the notable exception of N. Tranter, quoits are only mentioned in passing. N. Tranter, "The Chronology of Organized Sport in Nineteenth-Century Scotland: A Regional Study. I – Patterns," *IJHS* 7.2 (September, 1990), p.195; M.A. Speak, "Social Stratification and Participation in Sport in Mid Victorian England with Particular Reference to Lancaster, 1840–1870," in J.A. Mangan (ed.), *Pleasure, Profit, Proselytism. British Culture at Home and Abroad, 1700–1914* (London, 1988), p.46.

63 To reinforce the point, the only scholarly article on quoits is N. Tranter "Organized Sport and the Working Classes of Central Scotland, 1820–1900. The Neglected Sport of Quoiting," in R. Holt (ed.), *Sport and the Working Class in Modern Britain* (Manchester, 1990), Chap.3, pp.45–66.

64 A. Tomlinson, "Shifting Patterns of Working-Class Leisure. The Case of Knur and Spell," *Sociology of Sport Journal* 9.2 (1992), pp.192–206. In fact, knur and spell was played, on occasion, in the coalfield *MH*, 31 May 1879.

65 Tranter, "The Neglected Sport of Quoiting,"pp.45–66.

66 J. Benson, *British Coalminers in the Nineteenth Century: A Social History* (Dublin, 1980), p.161; for Yorkshire, Tomlinson, *Shifting Patterns of Working-Class Leisure*, p.196.

67 *MH*, 2 November 1872.

68 *BNWT*, 21 April 1888.

69 *MH*, 8 September 1888.

70 *MH*, 15 June 1895.
71 Data taken from various court records. Morpeth Petty Sessions, Minute Book, 1871–75; Registers of the Court of Summary Jurisdiction, Morpeth, 1889–94, 1900, 1910, Blyth, 1900, 1905, 1910, 1913, Bedlington, 1900, 1901, 1905, 1910, 1913; East Castle Ward Petty Sessions, 1894–95.
72 This represents the continuation of traditional practices and lies outside the changes of the 1880s contained in R. McKibbin, "Working-Class Gambling in Britain, 1880–1939," *Past and Present* 82 (February, 1979), pp.147–178.
73 Register of the Court of Summary Jurisdiction, Morpeth, 1910.
74 A fascinating view of the game was provided in the interviews of James Watson, Ashington and Bob Reay, New Hartley.
75 Derived from Register of the Court of Summary Jurisdiction, Morpeth, 1910.
76 *MH*, 7 October 1876.
77 *MH*, 31 July 1897.
78 *BNWT*, 19 February 1904.
79 Data from Morpeth Petty Sessions, Minute Book, 1871–75; County of Northumberland Register of the Court of Summary Jurisdiction, Morpeth, 1900–13, Blyth, Bedlington, 1900–13.

4 The transformation of traditional sports

1 *NDC*, 5 April 1901.
2 Database constructed from *NDC*, 1858–1914; *NWC*, 1850–1914; *MH*, 1854–1900, 1911–14; *BNWT*, 1870–1914.
3 Arcot Club was still playing in 1856 (*MH*, 23 August 1856).
4 *MH*, 18 April, 13 June, 4 July, 18 July, 8 August, 5 September, 1868, 27 June, 1 August, 26 September, 1874.
5 *MH*, 9 August 1881. Longhirst played in 1877 but the ground was not identified until 13 May 1882.
6 First record of games (*MH*, 6 June 1891). But in 1912 the club held its 25th annual ball (*MH*, 1912). This means the club was in existence in 1887.
7 The Bothal Club first played in May 1891 (*MH*, 31 May 1891). Four Samples played in 1892 (*BNWT*, 6 August 1892).
8 *NDC*, 1 February 1893. Another hall team played throughout much of the century – Meldon Park.
9 *MH*, 26 September 1874; *BNWT*, 20 July 1878; *NDC*, 28 April 1905.
10 (Backworth Percy) *BNWT*, November 1906; *NDC*, 28 April 1905; (Newsham) *NDC*, 2 August 1904; (Blyth) *BNWT*, 17 November 1883; (Seghill) *NDC*, 19 April 1898; (East Holywell) *NDC*, 2 June 1899.
11 *MH*, 24 June 1876.
12 *MH*, 24 June 1881. Players identified through 1881 Census.
13 *MH*, 9 July 1881. 1881 Census.
14 *MH*, 28 May 1887.
15 *BNWT*, 9 June, 23 June, 14 July, 21 July, 15 September 1883.
16 1891 Census.
17 *MH*, 1 July 1899; *NDC*, 2 June 1899.
18 Interestingly, out of the 17 miners who were interviewed in the 1970s the only one to mention cricket was from New Hartley, the home of Hartley Hastings.
19 *NDC*, 9 February 1892.
20 *Ashington Colliery Magazine* 14.5 (May, 1934), p.139.
21 For the development of league cricket elsewhere see Jeff Hill, "League Cricket in the North and Midlands, 1900–1940," in R. Holt (ed.) *Sport and the Working Class in Modern Britain* (Manchester 1990), Chap.7, p.121; K.A.P. Sandiford, "Cricket and Victorian Society," *Journal of Social History* (Winter, 1983), p.304; Jeffrey Hill,

"First-Class Cricket and The Leagues," *IJHS* 4.1 (May, 1987), p.70; J. Bale, *Sport and Place* (London, 1982), p.72; J. Williams, "Cricket," in T. Mason (ed.), *Sport in Britain* (Cambridge, 1989), Chap.4, p.139; D. Birley, *A Social History of English Cricket* (London, 1999), pp.151–152.

22 The lukewarm response to league cricket casts doubts on Birley's claim that working class sport "Had a momentum of its own with fierce local loyalties and an unhibited zest for competition *which led inexorably to leagues, cups, the cult of the spectator and gate money and professionalism.*" D. Birley, *Land of Sport and Glory, Sport and British Society, 1887–1910*" (Manchester, 1995), p.32. The history of cricket in East Northumberland certainly calls into question the inevitability of leagues. This is given support by the essential rejection of leagues in quoits, potshare bowling, and billiards.

23 This analysis is drawn from the sources in note 21.

24 *BNWT*, 14 January 1913; *MH*, 10 January 1913.

25 In fact, rowing was the most popular spectator sport during the 1860s and 1870s, the high point of professional rowing matches on the River Tyne.

26 For an excellent account of the funeral of Harry Clasper, see D. Clasper, *Harry Clasper: Hero of the North* (Gateshead, 1990), pp.62–69.

27 *NDC*, 22 August 1860; *MH*, 31 August 1861; *MH*, 16 August 1862.

28 *MH*, 16 August 1862.

29 Data for this history of James Boyd from *MH*, 30 April 1864, 4 March, 25 September 1865, 20 April, 1867, 2 December 1871, 9 November 1872, 5 April 1873, 13 December 1873, 9 May 1874, 3 October 1874, 5 December 1874; *NDC*, 28 February 1865, 19 March 1866, 2 May 1870, 26 August 1871, 28 August 1871.

30 Data from *NDC* and *MH*, 1874–80.

31 *MH*, 16 March 1878.

32 *MH*, 28 September 1878.

33 Data from *NDC, MH, BNWT*, 1866–1912.

34 *NDC*, 14 November 1910.

35 *MH*, 26 December 1913.

36 *MH*, 4 August 1888.

37 *MH*, 13 August 1887.

38 *NDC*, 13 September 1899.

39 *NDC*, 11 April 1900, 12 May 1902, 18 March 1905, 4 August 1908, 19 August 1910.

40 *MH*, 19 August 1910.

41 *NDC*, 26 February 1906.

42 *NDC*, 13 June 1900, 21 June 1902, 19 June 1903, 28 June 1905, 22 June 1906, 25 June 1907, 23 June 1908, 11 June 1909, 25 June 1912.

43 *NDC*, 17 June 1896.

44 These conclusions are drawn from N. Wigglesworth, *The Social History of English Rowing* (London, 1992). In particular these issues are addressed in chapters 3 ("Commercialism"), 4 ("Professionalism") and 6 ("Amateurism"). Additionally they are addressed in M.A. Speak, "Social Stratification in Sport in Mid-Victorian England with Particular Reference to Lancaster, 1840–1870," in J.A. Mangan (ed.), *Pleasure, Profit, Proselytism. British Culture and Sport at Home and Abroad, 1700–1914* (London, 1988), p.50; N. Wigglesworth, "A History of Rowing in the North West of England," 4.1 (September, 1986), pp.145–157; E. Halladay, "Of Pride and Prejudice. The Amateur Question in *British Journal of the History of Sport* (*BJHS*) English Nineteenth Century Rowing," *IJHS* 4.1 (May, 1987), pp.39–55.

45 The difference was clearly recognized in J. Mott, "Miners, Weavers and Pigeon Racing," in M. Smith, S. Parker and C. Smith (eds) *Leisure and Society in Britain* (London, 1973), p.86: "Short-distance racing, intensely communal, disreputable and associated with gambling . . . and long-distance racing, intensely competitive, national and very respectable."

46 From *NDC*, 1890 – January (7), February (2), April (7), May (15), June (10), July (6),

August (4), September (5), October (3), November (1); and MH, 1890 – February (1), May (4), July (2), September (2), October (1), November (1).

47 There were only two inns/public houses in Dudley.

48 Taped interview, Jimmy Tabarar.

49 The popularity of short distance flying was attested to by the fact that six of the miners interviewed talked about the short distance races.

50 Taped interview with Jack Dorgan.

51 The first reference to a club was in *MH*, 11 May 1889. However in 1908 they held the 17th Annual Meeting of Burradon and West Moor Homing Society (*NDC*, 9 November 1908). This would place the beginning at 1891.

52 *NDC*, 21 July 1892.

53 *MH*, 27 October 1900.

54 *NDC*, 7 October 1907.

55 *NDC*, 4 December 1907; *BNWT*, 29 November, 7 February 1910; *NDC*, 12 December 1910.

56 J. Benson, *British Coalminers in the Nineteenth Century. A Social History* (Dublin, 1980), p.162. There is virtually no writing on pigeon flying apart from Mott, *Miners, Weavers and Pigeon Racing*. This reflects graphically the huge holes in our understanding of the history of sport.

57 Hoppings were annual communal celebrations that were held in most villages during the summer. They were renowned for excessive drinking.

58 *Newcastle Journal*, 6 January 1844.

59 For a history of pedestrianism on Tyneside see F.C. Moffatt, *Turnpike Road to Tartan Track. A History of Professional Footrunning on Tyneside, 1850 to 1970* (Morpeth, 1979) and M. Kirkup, *The Biggest Mining Village in the World. A Social History of Ashington* (Morpeth, 1993), pp.142–143.

60 Interestingly, the only writing on the Morpeth Olympic Games has been from Germany. J. Thielgen, "Genesis and History of the Morpeth Olympic Games from 1873 to 1958," Unpublished "Staatsarbeit" in partial fulfilment of the requirements of the final examination for grammar school teachers. Also J. Ruhl, "The History and Development of the Morpeth Olympic Games (1871–1858)," Paper presented at ISHPES Congress, Budapest, 1999 (to be published in Proceedings).

61 *NDC*, 10 October 1870.

62 Moffatt, *Turnpike Road to Tartan Track*, pp.11, 23.

63 This was the result of a campaign, by the authorities, to close all commercial grounds in Northumberland. They were successful for a brief period, then new grounds opened up. Again demonstrating the limits of the power of the authorities.

64 *NDC*, 25 May 1895.

65 *MH*, 16 June 1860.

66 *MH*, 28 June 1879.

67 During the 1890s, an infrastructure of handicappers and starters developed. One of the most successful was Jos Tully. From the newspapers it is evident he was busy every weekend during the summer. At the Morpeth Olympic Games this often meant handicapping over 500 participants in the 120 yard handicap, the 110 yard handicap and the 300 yard handicap. Quite a task. Jos Tully was a working miner who died in a pit accident in 1920 (Moffatt, *Turnpike Road to Tartan Track*, p.63).

68 *NDC*, 4 August 1890.

69 *NDC*, 5 September 1910.

70 *NDC*, 19 August 1901.

71 *NDC*, 4 August 1913.

72 This emphasis on "amateur" athletes with professionals always being portrayed as "bad" is reflected in J. Crump, "Athletics," in T. Mason (ed.), *Sport in Britain* (Cambridge, 1989), Chap.2, pp.45–77.

73 See *NDC*, 17 March 1894, 19 November 1895, 14 March 1900, 2 October 1906, Even

then the North East lagged behind other areas. For example, the Northern Cross Country Association was established in 1882.

74 Birley, *Land of Sport and Glory*, p.64; Bale, *Sport and Place*, p.110; Crump, "Athletics," p.51; Tranter, "The Chronology of Organized Sport in Nineteenth Century Scotland I – Patterns," p.371.

75 This point cannot be over-emphasized. Physical violence was an essential part of daily life. They literally fought at the drop of a hat. It is impossible to understand mining life without understanding the importance of physical violence to the concept of masculinity. See J. Maguire, "Images of Manliness and Competing Ways of Living in Late Victorian and Edwardian Britain," *BJHS* 3.3 (December, 1986), pp.265–287.

76 Taped interview, Jimmy Tabarar.

77 Taped interview, James Wilson.

78 G. Halliwell, "Fifty Years On," *NWC*, 19 January 1895.

79 *BNWT*, 17 April 1906.

80 *MH*, 18 November 1899.

81 Metcalf illustrates how deeply some innkeepers were involved in sport both as participants and promoters. Not only did he have a "stable" of boxers but he also promoted pedestrian matches, whippet racing, rabbit coursing and shooting. He was an active participant in the Bedlington Ariel Bicycle Club and in long distance pigeon racing. He was a strong supporter of the Bedlington Hoppings. He was not atypical.

82 *NDC*, 23 April 1901.

83 Data from *NDC*, 1906–11.

84 The most complete view of boxing is in S. Shipley, "Boxing," in T. Mason (ed.), *Sport in Britain* (Cambridge, 1989), Chap.3, pp.78–115.

85 J. Benson, *British Coalminers*, p.163; Shipley, "Boxing," pp. 81, 83.

86 Shipley, "Boxing," p.91; N. Wigglesworth, *The Evolution of English Sport* (London, 1996), p.6.

5 New sports for new times

1 In fact, this was the second club in Seaton Sluice. The first held races in August, 1880 (*MH*, 14 August 1880). However, there is no record of any further competitions.

2 Blyth, Newbiggin, Bedlington and Cambois. Blyth, Newbiggin and Bedlington all had professionals and businessmen. Cambois, on the other hand, was a mining village.

3 *MH*, 10 December 1881, 4 March 1882, 4 July 1885, 28 November 1885.

4 *MH*, 24 January, 31 January, 14 March, 30 May 1884, 9 March 1885.

5 The first clubs were Morpeth Lawn Tennis Club, 1887 and Blyth Lawn Tennis Club. For the Seaton Delaval Club see *MH*, 5 January 1889.

6 *MH*, 7 April 1894. Report of 2nd year of Ashington Lawn Tennis Club.

7 *NDC*, 15 and 16 September 1911.

8 *NDC*, 5 April 1895.

9 *NDC*, 17 April 1896.

10 *NDC*, 27 July 1900.

11 *MH*, 18 June 1910.

12 *NDC*, 19 May 1905.

13 *NDC*, 5 June 1876.

14 *MH*, 8 September 1877.

15 *MH*, 11 September 1880.

16 *MH*, 14 May 1887.

17 *MH*, 23 July 1887.

18 *NDC*, 13 April 1892.

19 *MH*, 24 October 1891.

20 Data from "Northumberland, Durham, and Cleveland Club Directory," in J.H. Morrison (ed.), *The Northern Cyclist Pocket Guide* (1897) *MH*, 1888, 1889, 1890, 1891, 1892,

1892, 1894, 1895; *NDC*, 1889, 1890, 1891, 1892, 1893, 1894, 1895; *BNWT*, 1892; *Newcastle Daily Journal*, 1895.

21 *MH*, 12 March, 10 September 1887, 21 January 1888, 16 February, 23 November, 14 December 1889, 15 March 1890, 7 May, 28 May 1892, 6 May, 16 December 1893, 23 February 1895; *BNWT*, 14 May 1892; *NDC*, 11 February 1889, 24 July 1890, 6 March, 25 May, 22 June, 30 November 1891, 26 April, 9 May 1892, 7 March, 5 May, 21 August, 21 November 1893, 2 April, 12 April 1894.

22 *BNWT*, 24 May 1889.

23 *NDC*, 5 February 1891.

24 Interestingly the Newcastle Branch of the NCU was not reported in the *Morpeth Herald* until 1892, the time coalfield clubs began to join the NCU (*NDC*, 5 February 1891).

25 *MH*, 6 May 1893; *NDC*, 1 May 1893.

26 *NDC*, 22 June 1891.

27 *NDC*, 9 December 1893.

28 *MH*, 23 February 1895; *NDC*, 28 January, 18 February, 27 February 1895.

29 *MH*, 1 March 1890.

30 *NDC*, 25 May 1891.

31 *BNWT*, 4 June 1892.

32 *NDC*, 5 May 1893.

33 *MH*, 20 August 1898.

34 *NDC*, 29 June 1903.

35 *MH*, 11 September 1880.

36 Taped interview, Jack Dorgan.

37 The identification of the club members' occupations is from the 1891 Census, Seaton Delaval.

38 *NDC*, 25 May 1891.

39 The critical importance of control of space and the impact upon people subject to that control is developed in M. Oikawa, "Cartographies of Violence: Women, Memory, and the Subjects of the Internment," *Canadian Journal of Law and Society / Revue canadienne droit et société* 15.2 (2000), pp.39–69.

40 This was a very important shift. It appears that the changing rationale started in the mid-1880s and gradually spread throughout the district. It became universal in the early 1890s.

41 *MH*, 24 December 1892.

42 *NDC*, 26 January 1891.

43 Paralleling and perhaps related to the shift to education and entertainment was the fact that activities became nightly rather than predominantly at the weekend. This represented a very real change.

44 *NDC*, 29 October 1892.

45 *MH*, 3 December 1892.

46 *NDC*, 15 March 1894.

47 *MH*, 3 December 1892.

48 *NDC*, 18 January 1896.

49 *NDC*, 9 February 1898.

50 *NDC*, 26 January, 15 May 1895.

51 *NDC*, 9 January 1897.

52 *NDC*, 20 April 1903.

53 From "Club and Institute Journal," *NDC*, 6 February 1914.

54 *NDC*, 10 January 1896.

55 *NDC*, 9 February 1898.

56 The data was gathered from newspaper annual reports of income and expenditures of 16 Workmen's Institutes and Miners' Institutes from 1895 to 1913: 1895 (2), 1896 (1), 1898 (2), 1899 (1), 1902 (4), 1903 (2), 1907 (1), 1913 (2). These were contained in the *NDC* and *MH*.

57 Nine of the interviewed miners talked about playing billiards at the institute.
58 *NDC*, 15 May 1895.
59 *NDC*, 17 March 1907.
60 *NDC*, 20 March 1907.
61 *NDC*, 26 January, 10 April 1907.
62 *BNWT*, 20 February 1892.
63 *NDC*, 23 September 1897.
64 *NDC*, 20 November 1907.
65 *NDC*, 19 February 1896.
66 *NDC*, 20 February 1901.
67 *NDC*, 15 November 1905.
68 *NDC*, 2 February 1910.
69 *NDC*, 21 February, 24 April 1894.
70 *NDC*, 10 January 1899.
71 The first reference to Waterloo Rd. Billiard Saloon was in the *NDC*, 6 February 1896. The first reference to the Central Billiard Saloon was in the *NDC*, 3 December 1896.
72 Report of Annual Meeting, Northern Union of Mechanics' Institutes at Blyth (*MH*, 1 October 1898).
73 See N. Wigglesworth, *Evolution of English Sport* (London, 1996), pp. 39, 60, 80, 98, 114, 120; Thierry Terret, "Professional Swimming in England before the Rise of Amateurism, 1837–1875," *IJHS* 12.1 (April, 1995), pp.18–32.
74 The fullest coverage of angling is to be found in the work of John Lowerson. J. Lowerson, "Brothers of the Angle: Coarse Fishing and Working-Class Culture," in J.A. Mangan (ed.), *Pleasure, Profit, Proselytism* (London, 1988); J. Lowerson, "Angling," in T. Mason (ed.), *Sport in Britain: A Social History* (Cambridge, 1989), Chap.1, pp.12–43; N. Wigglesworth, *Evolution of English Sport*, p.68; D. Birley, *Land of Sport and Glory. Sport and British Society, 1887–1910* (Manchester, 1995), pp.27–29, 110. Birley calls "Fly fishing an elitist sport" But that was not totally the case in East Northumberland.
75 N. Tranter, "The Chronology of Organized Sport in Nineteenth Century Scotland: A Regional Study I – Patterns, *IJHS* 7.2 (September, 1990), p.189.
76 For golf clubs in Central Scotland see Tranter, *The Chronology of Organized Sport*, p.189; J. Lowerson, "Golf," in T. Mason (ed.), *Sport in Britain* (Cambridge, 1989), pp.187–214; D. Birley, *Land of Sport and Glory*, pp. 103–110; J. Lowerson, "'Scottish Croquet': The English Golf Boom, 1880–1914," *History Today* 33.5 (May, 1983), pp.25–30.
77 Tranter, *The Chronology of Organized Sport*, p.189; Birley, *Land of Sport and Glory*, p.202; H. Walker, "Lawn Tennis," in T. Mason, *Sport in Britain*, Chap.8, pp.243–275.
78 Even though there was a Northumberland and Durham Bowling Association in 1882, East Northumberland lagged far behind. See Birley, *Land of Sport and Glory*, p.89.
79 Tranter, *The Chronology of Organized Sport*. There has been no comprehensive history of cycling. The following provide insights into different elements: W. Vamplew, *Pay Up and Play the Game. Professional Sport in Britain, 1875–1914* (Cambridge, 1988), p.189; S. Creighton, "Organized Cycling and Politics: The 1890s and 1900s in Battersea," *The Sports Historian* 15 (May, 1995), pp.65–79.

6 Football: a symbol of sport in East Northumberland

1 This chapter is based on the research done for a paper on football. See A. Metcalfe, "Football in the Mining Communities of East Northumberland, 1882–1914," *IJHS* 5.3 (December, 1988), pp.269–291.
2 The history of football is well documented: T. Mason, *Association Football and English Society, 1863–1915* (Sussex, 1980); C. Korr, *West Ham United: The Making of a Football Club* (Chicago, 1986); S. Tischler, *Footballers and Businessmen: The Origins of Professional Soccer in England* (London, 1981); W. Vamplew, *Pay Up and Play the Game: Professional Sport in Britain, 1875–1914* (Cambridge, 1988); J. Walvin, *The People's Game: The History of Football Revisited*

(London, 1994). These have focused primarily on the high profile professional leagues. The actual playing of the game by the mass of society has received less treatment. See Metcalfe, "Football in the Mining Communities"; R. Holt, "Working Class Football and the City," *BJSH* 3.1 (May, 1986), pp.5–17.; M. Huggins, "The Spread of Association Football in North-East England, 1876–1890: The Pattern of Diffusion," *IJHS* 6.1 (December, 1989), pp.299–318; J. Walvin, *The People's Game*, Chap.3.

3 This supports Goulstone's contention that the "Local game was more advanced, more highly organized than that played by the chief footballing schools." J. Goulstone, "The Working Class Origins of Modern Football," *IJHS* 17.1 (March, 2000), p.135.

4 *NDC*, 7 January 1875.

5 *NDC*, 16 October 1882.

6 Data from *MH*, 1883–86. Number of teams: 1883 (7), 1884 (32), 1885 (62).

7 *Ibid.*

8 *MH*, 19 June, 31 July, 14 August, 11 September 1886.

9 *NDC*, 11 September 1913.

10 The antagonism between the Newcastle dominated NFA and the majority of teams in East Northumberland was one of the defining characteristics of the game until 1914. Antagonism towards Newcastle permeated all sport.

11 *MH*, 19 May 1888. Letter proposing formation of a "Tyne and Tweed Football Association." The name was changed to the Northern FA in August. The first annual meeting of the Northern Association was held in June 1889. At that meeting it was stated that the association was formed in April 1888 (*MH*, 15 June 1889).

12 *MH*, 12 October 1889.

13 This was one of the critical points in the history of the game in East Northumberland. They were forced to affiliate with the NFA if they wished to play any teams outside the coalfield. They approached the FA in February 1889 (*MH*, 2 March 1889). The FA refused to recognize the NA (*MH*, 12 October 1889).

14 The critical change from traditional to modern sport was the establishment of leagues and the development of the "season". This means that the critical change took place between 1893 and 1897.

15 Data compiled from reports of the NFA annual meetings, executive meetings and competitions, contained in the *NDC*, 1889–1914; *MH* 1889; *Newcastle Daily Journal* (*NDJ*) 1895, 1905; *BNWT* 1910.

16 M. Huggins, "The Spread of Association Football," p.311.

17 This is important. As the game developed deeper roots in the communities there developed multiple attachments – Newcastle United, Ashington FC, and Seaton Hirst. These teams played at different levels and thus one individual could legitimately support three or more teams.

18 The relationship between permanence and impermanence and financial viability is addressed in M. Huggins, "The Spread of Association Football," p.303; N. Tranter, "The Chronology of Organized Sport in Nineteenth Century Scotland: A Regional Study II – Causes," *IJHS* 7.3 (December, 1990), p.376.

19 *NDC*, 19 September 1901.

20 *MH*, 11 November 1893.

21 *NDC*, 10 November, 8 December 1896.

22 *NDC*, 20 May 1901.

23 *NDC*, 17 August 1909.

24 *BNWT*, 8 April 1913.

25 The history of Shankhouse encapsulates the history of small teams in East Northumberland and, I suggest, other areas, better than anything else.

26 *NWC*, 27 April 1895.

27 1891 Census.

28 *MH*, 12 December 1896; *NDC*, 10 November 1896.

29 *MH*, 5 June 1897.

30 *NDC*, 5 August 1896.
31 *MH*, 11 June 1898.
32 *NDC*, 31 October 1903.
33 *NDC*, 12 May 1906.
34 *NDC*, 27 May 1907.
35 Annual Meeting, Bedlington United FC (*BNWT*, 14 June 1910).
36 *NDC*, 12 July 1913.
37 *NDC*, 15 February 1895.
38 *MH*, 5 June 1897.
39 At the same time there were two amateur leagues in Newcastle.
40 *NDC*, 16 August 1906; *BNWT*, 17 August 1906.
41 *NDC*, 17 January 1893.
42 *MH*, 2 February 1895.
43 Minute Book, Earsdon UDC, 1907–1910, 17 August 1909.
44 *NDC*, 23 April 1896.
45 *MH*, 8 August 1896.
46 *BNWT*, 4 September 1906.
47 For the role of the "pubs" see M.A. Speak, "Social Stratification in Sport in Mid-Victorian England with Particular Reference to Lancaster, 1840–1870," in J.A. Mangan (ed.), *Pleasure, Profit, Proselytism. British Culture and Sport at Home and Abroad, 1700–1914* (London, 1988), p.50; R. Holt, "Working Class Football and the City," *BJSH* 3.1 (May, 1986), p.9; Mason, *Association Football*, pp.26–27; D. Birley, *Land of Sport and Glory: Sport and British Society, 1887–1910* (Manchester, 1995), p.35; J. Goulstone,"The Working Class Origins of Modern Football," *IJHS* 17.1 (March, 2000), p.139; Walvin, *The People's Game*, p.60.
48 There were 106 reports of annual meetings between 1891 and 1914.
49 *NDC*, 28 February 1898.
50 *BNWT*, 25 August 1913.
51 *NDJ*, 9 November, 12 December 1895; *NDC*, 6 January 1895.
52 *NDC*, 2 October 1895.
53 *NDC*, 11 October 1902.
54 *NDC*, 1 September 1905.
55 *NDC*, 24 January 1898.
56 *NDC*, 23 June 1913.
57 *NDC*, 1 January 1889.
58 The data was generated from the NFA annual reports and the frequent executive meetings that dealt with infringements. These reports were in the *Newcastle Journal* and *NDC*.
59 *MH*, 18 April 1885.
60 *NDC*, 18 April 1892.
61 *NDJ*, 6 February 1895.
62 *NDC*, 10 March 1896.
63 *Sporting Man*, 11 February 1901.
64 *Sporting Man*, 21 October 1901.
65 *NDC*, 30 May 1904.
66 *NDJ*, 16 October 1895.
67 Twenty-seventh annual report of NFA (NDC, 23 May 1910).
68 Report of NFA, Earsdon Ground closed from 13 December (*NDC*, 10 December 1909). Monthly Council meeting of NFA (*NDC*, 11 February 1910).
69 Photo of Barrington Girls' Football Team, 1921. They played 12 other teams all located in the northern part of the coalfield. S. Martin, *Barrington Colliery Village* (Bedlington, 1979), p.25.
70 For the development of associations see Mason, *Association Football*, p.15.
71 Mason, *Association Football*, p.85. Dates of Associations: London, 1885; Manchester, 1891; Liverpool, 1891; Nottingham, 1891; Brighton, 1892; Sunderland, 1893; Leicester, 1893.

72 M. Huggins, "The Spread of Association Football, p.305.

73 For churches see Walvin, *The People's Game*, p.59; Mason, *Association Football*, p.25; S. Barlow, "The Diffusion of Rugby Football in the Industrialized Context of Rochdale, 1868–1890: A Conflict of Ethical Values," *IJHS* 10.1 (April, 1993), p.54. For the workplace see Mason, *Association Football*, pp.28–29; R. Holt, "Working Class Football in the City," p.7.

Conclusion

1 Taped interview with Jimmy Tabarar.

2 J. Williams, "Churches, Sport and Identities in the North, 1900–1939," in J. Hill and J. Williams (eds), *Sport and Identity in the North of England* (Keele, 1996), Chap.6, p.132.

3 M. Huggins, "Second-class Citizens? English Middle-class Culture and Sport, 1850–1910: A Reconsideration," *IJHS* 17.1 (March, 2000), p.6.

References

Primary

General

Address to the Pitmen, Keelmen, and Other Labouring Classes Employed on the Tyne and Wear (Durham, 1819).

Bell, J.G., *Collections Forming a General History of Coal, Colliery-Engineering, and Mining, Together with The Local History of the Collieries and Coal Trade of the North of England*, 22 Vols (London, n.d.).

Colliers of the United Association of Durham and Northumberland, *A Voice from the Coal Mines, or a Plain Statement of the Various Grievances of the Pitmen of the Tyne and Wear* (South Shields, 1825).

Forbes, R., *An Essay on the Abridgement of the Hours of Labour in Mines* (Coatbridge, 1844).

Latimer, J., *Local Records, or Historical Register of Remarkable Events which have occurred in Northumberland, Durham, Newcastle-Upon-Tyne and Berwick-Upon-Tweed, 1832–1857* (Newcastle, 1857).

Liddell, D., *Suggestions Relative to the Best Means for Diffusing Useful Knowledge Among the Miners in The North of England* (Newcastle, 1836).

Rules and Regulations of the Coal Miners Friendly Society, in the Counties of Northumberland and Durham (Newcastle, 1837).

Rules and Regulations of the Miners' Association (London, 1842).

Sport and leisure

Sport and leisure

Bedlington Brass and String Band, Minute Book, 1890–99.

Blyth Recreation Company Ltd. Directors Minute Book, 1893–97, 1898–1907.

Laski-Holmes Letters, September 1924, September 1925.

Morpeth Harriers FC, Minute Book, 1884–90.

Morrison, J.H. (ed.), *Northern Cyclists Pocket Guide*, (1897).

—— *Northern Cyclists Pocket Guide*, (1901).

National Cyclists Union, Newcastle Centre, Accounts and Correspondence, 1903–10.

Northumbrian Anglers' Federation, *Handbook and Guide*, 1957.

Williamson, J. (ed.) *Life, Death and Burial of James Renforth, Champion Oarsman*, (1871). (A collection of newspaper cuttings.)

Miner poets

Allen's Illustrated Edition of Tyneside Songs and Readings (Newcastle, 1891).
Corven, E., *Random Rhymes, Being a Collection of Local Songs and Ballads* (Newcastle, 1850).
Skipsey, J., *Carols from the Coal Fields* (London, 1886).
Tate, M., *Poems, Songs and Ballads* (Blyth, 1898).
Wilson, T., *The Pitman's Pay and Other Poems* (London, 1872).

Papers

Sample Papers

Ashington and District Working Men's Social Club Ltd. Statement of Accounts, 1 July–29 November 1902.
Biographical material on Sample family, 10 items.
Bothal Show programme, 1889–1914.
Correspondence between Thomas Sample and William Thompson, Secretary of Committee, re Ashington Recreation Ground and Drill Hall, 13 letters, 14 October 1886–20 June 1889.
Correspondence re Ashington and District Working Men's Social Club.
Correspondence re Bothal footpath case, 1874–76.
Correspondence re keeping of dogs in Ashington and Pegswood, 10 letters, 1 September 1885–20 September 1886.
Correspondence re proposed angling association, 1860 (3 letters), 1892 (4 letters).
Correspondence re proposed new Picture Hall, 1913–14.
Correspondence re Shooting Rights, April 1884–April 1885, 1885–89, 15 August 1907, 8 October 1913–3 December 1913.
Correspondence, William Ellis to Sample, re Pegswood Club, 4 January 1904.
Diaries of Thomas Sample, 1902.
Diary of W.C. Sample, 1913.
Prospectus of Dudley and Weetslade Social Club Ltd., n.d.
Game Books, 1901, 1908–09, 1913–14.
Memorandum and Articles of Association of the Ashington and District Working Men's Social Club Ltd, 18 January 1902.

Ridley Papers

Correspondence between Sir M.W. Ridley and Blyth Council re use of links, 1877–78, 1907–08.
Diary of Sir M.W. Ridley, Hounds, 1830–33.
Document concerned with the plans for the commercial development of Sir M.W. Ridley's estates in South Blyth and Newsham, 1909–10.
Game Books, Blagdon, 1864–75.
Game Books, Blagdon, 1905–14.
Package of letters re public park question at Blyth, 1895–97.

Records

School records

Hirst North County Primary School, Log Book, 1896–August 1905.
Hirst North School, Log Book, 1896–1905.
Inspectors' Report, Cowpen Crofton Council School, 1904.
Minute Book, Bedlington, Barrington Colliery and West Sleekburn County Primary School, 1905–40.
Minute Book, Blyth Secondary School, 1905–13.
Minute Book of Managers, Bedlington Council School, 1908–.
Minute Book of Managers of Choppington Colliery Council Schools, 1904–22.
Nineteenth Century Board Schools in Northumberland, 1902.
North Seaton Mixed School, Log Book, 1885–1914, 1905–.
Old Hartley Council School, Log Book, 1875–1907.
School Correspondents Record Book for Schools in Seaton Delaval area, 1902.
School Correspondents Record Book for Schools in Seaton Delaval Area, 1903– .
Seaton Delaval English Presbyterian School, 1902–09.
Seaton Terrace Junior School, Log Book, 1909–29.
Seghill Infants School, Log Book, 1876–1903.

Church/chapel records

Bates Cottages Methodist Church Centenary Handbook, 1868–1968.
Bedlington Presbyterian Church, Minute Book, 1907–.
Blyth Congregational Church, Minute Book, 1904–13.
Blyth Parish Church Magazine, 1892–94, 1895, 1896, 1900–01, 1902–03.
Blyth Station, Primitive Methodist Connexion, Circuit Schedules, 1870–80.
Blyth Wesleyan Methodist Circuit, Schedule Book, 1841–47, 1847–73.
Choppington Primitive Methodist Church, Treasurers Book, 1895–1946.
Guide Post Methodist Church, A Memoir, n.d.
Primitive Methodist Connexion, Blyth Circuit, Sunday School Union, Minute Book, 1894–1903.
Primitive Methodist Connexion, Seaton Delaval Circuit, Delaval House Committee, Minute, 1882–97.
Seaton Delaval and New Hartley Magazine, 1912–15.
Seaton Delaval Primitive Methodist Circuit, Quarterly Meetings, 1875–1902.
Seaton Delaval Primitive Methodist Circuit, Station Reports, Chapel Schedules, Sunday School Returns, 1874–83, 1888–93.
Seaton Hirst Mission, Minutes of Trustees, 1902–71.
Seaton Hirst Primitive Methodist Mission, Minute Book of Leaders and Society Meetings, 1902–66.
Seghill Wesleyan Reform, Sunday School Reports, 1854–60.
Seghill Wesleyan/Wesleyan Reform/United Methodist Free Church, Sunday School Report, June 1854–January 1861.
Seghill Parish Magazine, 1902.

Court records

East Castle Ward Petty Sessions, 1894–95.
East Castle Ward, Petty Sessional Division, Register of Clubs, 1903–14.
Licensing Act, 1872, Division of Bedlingtonshire, Register of Licences, 1872–1904.
Morpeth Petty Sessions, Minute Book, 1845–49, 1850, 1851, 1861, 1862, 1864, 1871–75.
Morpeth Petty Sessions, Register of the Court of Summary Jurisdiction, 1889–94.
Morpeth, Register of the Court of Summary Jurisdiction, 1900, 1910.
Morpeth Ward, Petty Sessional Division, Register of Clubs, 1901–14.
Morpeth Ward, Register of Licenses, 1872–1900.
Morpeth Ward, Summary Jurisdiction, 1879.
Northumberland, Register of the Court of Summary Jurisdiction, Blyth and Bedlington, 1900–01, 1905, 1910, 1913.

Government

Census of Great Britain, *Population Tables. Number of the Inhabitants in the Years 1801, 1811, 1821, 1831, 1841, 1851*, Vol. II (London, 1852)
—— *Seaton Delaval, Holywell*, 1841, 1851, 1861.
—— *Seaton Delaval, Holywell, Bates Cottage, New Hartley, Seghill, Shankhouse, East Cramlington*, 1871, 1881.
—— *Seaton Delaval, Bates Cottages, Holywell, West Holywell, East Holywell, New Hartley, Seghill, Seaton Sluice, Seaton Terrace, Cramlington, Shankhouse, East Cramlington*, 1891.
Report of the Commissioners of Children's Employment Commission. Mines, Vol. 10 (London, 1842).
Report of the Commissioners on the Mining Districts, 1847, 1850, 1854.
Report of the Commissioners to Inquire into the (Operation of That Act) State of the Population in the Mining Districts, 1846 (London, 1846).
Report of the Inspector of Mines, 1864.
Report of the Select Committee of the House of Lords. The State of the Coal Trade in the United Kingdom. Evidence from Northumberland and Durham, 1829 (Newcastle, 1829).

Local government

Minute Book, Bedlingtonshire Urban District Council, 21 April 1893–8 April 1897, 1897–1900, 1910–1911, 1913–1914.
Minute Book, Blyth Urban District Council, April 1907–January 1909, 11 February 1909–7 June 1910, July 1910–November 1911, December 1911–June 1913, July 1913–April 1915.
Minute Book, Cowpen District Local Board, 22 December 1875–11 February 1880, 22 October 1889–12 November 1891.
Minute Book, Cowpen Urban District Council, 13 June 1895–10 December 1896, 12 January 1897–8 September 1898, 8 September 1898–12 February 1901, 14 February 1901–14 May 1903, 14 May 1903–4 August 1905, August 1905–March 1907.
Minute Book, Cramlington Urban District Council, 1903–08; 4 May 1908–February 1911, 14 March 1911–May 1913.
Minute Book, Earsdon Urban District Council, 1903–07, 1907–10, 1910–13.
Minute Book, Newbiggin-By-The-Sea, January 1894–March 1898, March 1898–March 1902.

Minute Book, Newbiggin Urban District Council, 2 April 1902–December 1905, January 1906–February 1909, March 1909–January 1913.
Minute Book, Seaton Delaval Parish Council, 1895–1902, 1902–08.
Minute Book, Seaton Delaval Urban District Council, 1912– 14.
Minute Book, Seghill Urban District Council, 12 January 1905–8 October 1914.
Minute Book, South Blyth Urban District Council, 18 August 1896–11 January 1899, 24 January 1899–5 March 1902, 12 March 1902–2 February 1905, 3 February 1905–9 December 1907.

Other

Directories

Kelly's Directories of Northumberland and Durham, 1858, 1873, 1887, 1894, 1902, 1910, 1914.
Parson, and White, W. *History, Directory and Gazetter of the Counties of Durham and Northumberland And the Towns and Counties of Newcastle-Upon-Tyne*, 2 vols (Leeds, 1827).
Robson's Directory of Northumberland, 1841.
Slater's Directory of the Northern Counties, Vol. 1, 1855.
Ward's North of England Directories, 1854, 1855.

Other

Ashington Collieries Deputations Book, 1893–96.
Fordyce, T., *Local Records or Historical Register* (Newcastle, 1867).
Forster Papers.
John Watson Papers.
Memoir of John George Weeks. Read before the North of England Institute of Mining and Mechanical Engineers by R.J. Weeks, 9 February 1918.
Morpeth Collectanea, Vol. IV.
Seaton Delaval and New Hartley Conventional Magazine, 1913–15.
Records of John Swann, farmer, Lookout Farm, Seaton Delaval Hall, 1891–1912.
Stamp, Richard, *Recollections of Richard Stamp* (Newcastle, 1891).
Tapes of interviews with 21 miners/miners' wives in Northumberland Record Office.
Ward, Mr, *A Selection of Notes on Ashington* (Ashington, n.d.)

Newspapers, journals and magazines

Blyth News and Weekly Transcript, 1870–1914.
Morpeth Herald, 1854–1900, 1911–14.
Newcastle Courant, 1848, 1850, 1851.
Newcastle Daily Chronicle, 1858–1914.
Newcastle Journal, 1840, 1885.
Newcastle Weekly Chronicle, 1821–23, 1831, 1840, 1844, 1850–1914.
The Ashington Collieries Magazine, 5.1 (January, 1925), 5.2 (February, 1925), 9.5 (May, 1929), 14.5 (May, 1934), 15.4 (April, 1935), 18.3 (March, 1938), 20.4 (April, 1940).
The Monthly Chronicle of North Country Lore and Legend, 1887–91.
The Northerner and Athlete, 3.83 (8 February 1893), 3.87 (8 March 1893), 3.90 (29 March 1893).

Secondary

Sport/Leisure

British sport history

Bale, J., *Sport and Place* (London, 1982).
Birley, D., *A Social History of English Cricket* (London, 1999).
—— *Land of Sport and Glory. Sport and British Society, 1887–1910* (Manchester, 1995).
—— *Sport and the Making of Britain* (Manchester, 1993).
Clasper, D., *Harry Clasper: Hero of the North* (Newcastle, 1990).
Cleaver, H., *A History of Rowing* (London, 1957).
Fairfax-Blakeborough, J., *Northern Turf History*, Vol. II (1949).
Hill, J. and Williams, J. (eds) *Sport and Identity in the North of England* (Keele, 1996).
Holt, R. *Sport and the British: A Modern History* (Oxford, 1989).
—— (ed.), *Sport and the Working Class in Modern Britain* (Manchester, 1990).
Huggins, M., *Flat Racing and British Society, 1790–1914: A Social and Economic History* (London, 2000).
Itzkowitz, D.C., *Peculiar Privilege: A Social History of English Fox Hunting, 1753–1885* (Sussex, 1977).
Jones, S.G., *Sport, Politics and the Working Class* (Manchester, 1988).
Korr, C., *West Ham United: The Making of a Football Club* (Chicago, 1986).
Lowerson, J., *Sport and the English Middle Classes, 1870–1914* (Manchester, 1993).
McCrone, K., *Sport and the Physical Emancipation of English Women, 1870–1914* (London, 1988).
Mangan, J.A., *Athleticism in the Victorian and Edwardian Public School* (Cambridge, 1981; London, 2000).
—— (ed.), *Pleasure, Profit, Proselytism. British Culture and Sport at Home and Abroad, 1700–1914* (London, 1988).
Mason, T., *Association Football and English Society, 1863–1915* (Sussex, 1980).
—— *Sport in Britain* (London, 1988).
—— (ed.), *Sport in Britain: A Social History* (Cambridge, 1989).
Strutt, J., *The Sports and Pastimes of the People of England*, edited by W. Hone (London, 1898).
Tischler, S., *Footballers and Businessmen: The Origins of Professional Soccer in England* (New York, 1981).
Vamplew, W., *Pay Up and Play the Game: Professional Sport in Britain, 1875–1914* (Cambridge, 1988).
—— *The Turf: A Social and Economic History of Horse Racing* (London, 1976).
Wagg, S., *The Football World* (Brighton, 1984).
Walvin, J., *Football and the Decline of Britain* (Macmillan, 1986).
—— *The People's Game: A Social History of British Football* (Arrow, 1975).
—— *The People's Game: The History of Football Revisited* (Mainstream Publishing, 1994).
Wigglesworth, N., *The Evolution of British Sport* (London, 1996).
—— *The Social History of English Rowing* (London, 1992).
Woodforde, J., *The Story of the Bicycle* (London, 1980).

Leisure

Bailey, P., *Leisure and Class in Victorian England: Rational Recreation and the Contest for Control, 1830–1885* (London, 1978).

Clarke, J. and Critcher, C., *The Devil Makes Work. Leisure in Capitalist Britain* (Chicago, 1985).

Cunningham, H., *Leisure and Culture* (London, 1990).

—— *Leisure and the Industrial Revolution, c.1780–c.1880* (London, 1980).

De Grazia, S., *Of Time, Work and Leisure* (New York, 1962).

Deem, R., *All Work and No Play? The Sociology of Women and Leisure* (Open University Press, 1986).

Hargreaves, J. (ed.), *Sport, Culture and Ideology* (London, 1982).

Harrison, B., *Drink and the Victorians: The Temperance Question in England, 1815–1872* (Pittsburgh, 1971).

Lowerson, J. and Myerscough, J., *Time to Spare in Victorian England* (Sussex, 1977).

Malcolmson, R.W., *Popular Recreations in English Society, 1700–1850* (Cambridge, 1973).

Meller, H.E., *Leisure and The Changing City, 1870–1914* (London, 1976).

Pimlott, J.A.R., *The Englishman's Holiday* (London, 1947).

Rose, J., *The Intellectual Life of the British Working Classes* (London, 2001).

Smith, M., Parker, S. and Smith, C.S. (eds), *Leisure and Society in Britain* (Allen Lane, 1973).

Walton, J.K., *The English Seaside Resort. A Social History, 1750–1914* (Leicester, 1983).

Walvin, J., *Leisure and Society, 1830–1950* (London, 1978).

Chapters in books

Crump, J., "Athletics," in T. Mason (ed.), *Sport in Britain. A Social History* (Cambridge, 1989), Chap.2, pp.44–77.

Cunningham, H., "Leisure," in J. Benson (ed.), *The Working Class in England, 1875–1914* (London, 1985), Chap.5, pp.133–164.

—— "Leisure and Culture," in F.M.L. Thompson (ed.), *The Cambridge Social History of Britain, 1750–1950. Vol. 2 People and Their Environment* (Cambridge, 1990), Chap.6, pp.279–339.

Hill, J., "League Cricket in the North and Midlands, 1900–1940," in R. Holt (ed.), *Sport and the Working Class in Modern Britain* (Manchester, 1990), Chap.7, pp.121–141.

—— "Rite of Spring: Cup Finals and Community in the North of England," in J. Hill and J. Williams (eds), *Sport and Identity in the North of England* (Keele, 1996), Chap.5, pp.85–112.

Holt, R., "Heroes of the North: Sport and the Shaping of Regional Identity," in J. Hill and J. Williams (eds), *Sport and Identity*, Chap.7, pp.137–164.

Lowerson, J., "Angling," in T. Mason (ed.), *Sport in Britain* (Cambridge, 1989), Chap.1, pp.12–43.

—— "Golf," in T. Mason (ed.), *Sport in Britain* Chap.6, pp. 187–214.

Mason, T., "Football, Sport of the North," in J. Hill and J. Williams (eds), *Sport and Identity*, Chap.3, pp.41–52.

Metcalfe, A., "'Potshare Bowling' in the Mining Communities of East Northumberland, 1800–1914," in R. Holt. (ed.), *Sport and the Working Class in Modern Britain* (Manchester, 1990), Chap.2, pp.29–44.

—— "Sport and Community: A Case Study of the Mining Villages of East Northumberland, 1800–1914," in J. Hill and J. Williams (eds), *Sport and Identity*, Chap.2, pp.13–40.

Morris, R.J. "Clubs, Societies and Associations," in F.M.L. Thompson, *The Cambridge Social History of Britain, 1750–1950. Vol. 3, Social Agencies and Institutions* (Cambridge, 1990), Chap.8, pp.395–443.

Parratt, C.M., "The Making of the Healthy and Happy Home: Recreation, Education, and the Production of Working-class Womanhood at the Rowntree Cocoa Works, York, c.1898–1914," in J. Hill and J. Williams (eds), *Sport and Identity*, Chap.4, pp.53–84.

Reid, D.A., "Beasts and Brutes: Popular Blood Sports c.1780–1860," in R. Holt (ed.), *Sport and the Working Class*, Chap.1, pp.12–28.

Shipley, S., "Boxing," in T. Mason (ed.), *Sport in Britain* Chap.3, pp.78–115.

Tranter, N.L., "Organized Sport and the Working Classes in Central Scotland, 1820–1900: The Neglected Sport of Quoiting," in R. Holt (ed.), *Sport and the Working Class*, Chap.3, pp.45–66.

Walker, H., "Lawn Tennis," in T. Mason (ed.), *Sport in Britain*, Chap.8, pp.245–275.

Williams, J., "Churches, Sport and Identities in the North, 1900–1939," in J. Hill and J. Williams (eds), *Sport and Identity*, Chap.6, pp.113–136.

Articles

Arnold, A.J., "The Belated Entry of Professional Soccer into the West Riding Textile District of Northern England: Commercial Imperatives and Problems," *IJHS*, 6.3 (December, 1989), pp. 319–334.

Bacon, W. and Pitchford, A., "Sports Development and the Public Sector: The Case of Association Bowls," *IJHS* 17 (March, 2000), pp. 165–175.

Bailey, P., "A Mingled Mass of Perfectly Legitimate Pleasures of the Victorian Middle Class and the Problem of Leisure," *Victorian Studies* 21.1 (Autumn, 1977), pp. 7–28.

—— "'Will the real Bill Banks stand up?' Towards a Role Analysis of Mid-Victorian Working-Class Respectability," *Journal of Social History* 12.2 (Spring, 1979), pp. 336–353.

Barlow, S., "The Diffusion of 'Rugby' Football in the Industrialized Context of Rochdale, 1868–90: A Conflict of Ethical Values," *IJHS* 10.1 (April, 1993), pp. 49–67.

Collins, T., "'Noa Mutton, Noa Laaking': The Origins of Payment of Play in Rugby Football, 1877–86," *IJHS* 12.1 (April, 1995), pp. 33–50.

Creigton, S., "Organised Cycling and Politics: The 1890s and 1900s in Battersea," *The Sports Historian* 15 (May, 1995), pp. 65–79.

Durie, A.J. and Huggins, M.J., "Sport, Social Tone and the Seaside Resorts in Great Britain, c.1850–1914," *IJHS* pp. 173–186.

Fletcher, S., "The Making and Breaking of a Female Tradition: Women's Physical Education in England 1880–1980," 2.1 (May, 1985), pp. 29–39.

Goulstone, J., "The Working-Class Origins of Modern Football," *IJHS* 17 (March, 2000), pp. 135–143.

Greenhalgh, P., "'The Work and Play Principle': The Professional Regulations of the Northern Rugby Football Union, 1898–1905," *IJHS* 9.3 (December, 1992), pp.356–377.

Halladay, E., "Of Pride and Prejudice: The Amateur Question in English Nineteenth-Century Rowing," *IJHS* 4.1 (May, 1987), pp. 39–55.

Hargreaves, J.A., "Playing Like Gentlemen While Behaving Like Ladies: Contradictory Features of the Formative Years of Women's Sport," *BJHS* 2.1 (May, 1985), pp.40–52.

Harrison, B., "Religion and Recreation," *Past and Present* 30 (1967), pp.98–125.

Hill, J., "First Class Cricket and the Leagues: Some Notes on the Development of English Cricket, 1900–40," *IJHS* 4.1 (May, 1987), pp.68–81.

Holt, R.J., "Cricket and Englishness: The Batsman as Hero," *IJHS* 13.1 (March, 1996), pp.48–70.

—— "Working Class Football and the City," *British Journal of Sports History* 3.1 (May, 1986), pp.5–17.

Howard, S., "Leisure in the Pit Village: Meaning and Change," *North East Labour History Bulletin* 27 (1993), pp.14–23.

Huggins, M., "Culture, Class and Respectability: Racing and the English Middle Classes in the Nineteenth Century," *IJHS* 11.1 (April, 1994), pp.19–41.

—— "Mingled Pleasure and Speculation: The Survival of Enclosed Racecourses on Teeside, 1885–1902," *British Journal of Sports History* 3.2 (September, 1986), pp.158–172.

—— "More Sinful Pleasures. Leisure, Respectability and the Male Middle Class in Victorian England," *Journal of Social History* 33.3 (Spring, 2000), pp.585–606.

—— "Second-Class Citizens? English Middle-Class Culture and Sport 1850–1910: A Reconsideration," *IJHS* 17 (March, 2000), pp.1–35.

—— "The Spread of Association Football in North-East England, 1876–90: The Pattern of Diffusion," *IJHS* 6.1 (December, 1989), pp.299–318.

Jones, G.S., "Working-Class Culture and Working-Class Politics in London, 1870–1900: Notes on the Remaking of a Working Class," *Journal of Social History* 7 (1974), pp.460–508.

Joyce, P., "The Return of History: Postmodernism and the Politics of Academic History in Britain," *Past and Present* 158 (February, 1998), pp.207–235.

Kellett., M.A., "The Power of Princely Patronage: Pigeon Shooting in Victorian England," *IJHS* 11.1 (April, 1994), pp.63–85.

Lewis, R.W., "Football Hooliganism in England Before 1914: A Critique of the Dunning Thesis," *IJHS* 13.2 (December, 1996), pp.310–339.

Lowerson, J., "'Scottish Croquet': The English Golf Boom, 1880–1914," *History Today* 33.5 (May, 1983), pp.25–30.

—— "Sport and the Victorian Sunday: The Beginnings of Middle-class Apostasy," *The British Journal of Sports History*, 1.2 (September, 1984).

—— "Sporting Metaphors and New Marathons: The Vitality of the Victorian Middle-Class Legacy," *IJHS* 17.4 (December, 2000), pp.111–122.

McKibbin, R., "Working-Class Gambling in Britain 1880–1939," *Past and Present* 82 (February, 1979), pp.147–178.

Maguire, J., "Images of Manliness and Competing Ways of Living in Late Victorian and Edwardian Britain," *British Journal of Sports History* 3.3 (December, 1986), pp.265–287.

Mangan, J.A., "The Social Construction of Victorian Femininity: Emancipation, Education and Exercise," *IJHS* 6.1 (May, 1989), pp.1–9.

Melling, J., "Sport, Spectacle and Class: Handball in the North East," *IJHS* 7.3 (December, 1990), pp.443–447.

Metcalfe, A., "Football in the Mining Communities of East Northumberland, 1882–1914," *IJHS* 5.3 (December, 1988), pp.269–291.

—— "Organised Sport in the Mining Communities of South Northumberland, 1800–1889," *Victorian Studies* 25.4 (Summer, 1982), pp.469–495.

—— "Resistance to Change: Folk Games in the Mining Communities of East Northumberland, 1800–1914," *Stadion* 15 (1989), pp.143–149.

—— "Sport and Space: A Case Study of the Growth of Recreational Facilities in East Northumberland, 1850–1914," *IJHS* 7.3 (December, 1990), pp.348–364.

—— "The Control of Space and the Development of Sport," *The Sports Historian* 15 (May, 1995), pp.3–22.

—— "The Development of Sporting Facilities: A Case Study of East Northumberland, England, 1850–1914," *International Review for the Sociology of Sport* 28 (1993), pp.107–119.

Munting, R., "Social Opposition to Gambling in Britain: An Historical Overview," *IJHS* 10.3 (December, 1993), pp.295–312.

Murphy, P., Dunning E., and Maguire, J., "Football Spectator Violence and Disorder before the First World War: A Reply to R.W. Lewis," *IJHS* 15.1 (April, 1998), pp.141–161.

Oikawa, M., "Cartographies of Violence: Women, Memory, and the Subjects of the Internment," *Canadian Journal of Law and Society* 15.2 (2000), pp.39–69.

Park, J., "Sport, Dress Reform and the Emancipation of Women in Victorian England: A Reappraisal," *IJHS* 6.1 (May, 1989), pp.10–30.

Parratt, C., "Athletic Womanhood: Exploring Sources for Female Sport in Victorian and Edwardian England," *Journal of Sport History* 16.2 (Summer, 1989), pp.140–157.

—— "Working-Class Women and Leisure in Late Victorian and Edwardian England," *IJHS* 15.2 (August, 1998), pp.22–53.

Ruhl, J., "The History and Development of the Morpeth Olympic Games (1871–1958)," Paper presented At ISHPES Congress, Budapest, 1999.

Russell, D., "'Sporadic and Curious': The Emergence of Rugby and Soccer Zones in Yorkshire and Lancashire c.1860–1914," *IJHS* 5.2 (September, 1988), pp.185–205.

Sandiford, K., "Cricket and the Victorian Society," *Journal of Social History* 17 (Winter, 1983), pp.303–317.

—— "Sport and Victorian England," *Canadian Journal of History* 18.1 (April, 1983), pp.111–117.

Terret, T., "Professional Swimming in England Before the Rise of Amateurism, 1837–75," *IJHS* 12.1 (April, 1995), pp.18–32.

Tomlinson, A., "Shifting Patterns of Working Class Leisure: The Case of Knurr and Spell," *Sociology of Sport Journal* 9.2 (June, 1992), pp.192–206.

Tranter, N., "Organized Sport and the Middle Class Woman in Nineteenth Century Scotland," *IJHS* 6.1 (May, 1989), pp.31–48.

—— "Popular Sports and the Industrial Revolution in Scotland: The Evidence of Statistical Accounts," *IJHS* 4.1 (May, 1987), pp.21–38.

—— "The Chronology of Organized Sport in Nineteenth Century Scotland: A Regional Study I – Patterns," *IJHS* 7.2 (September, 1990), pp.188–203.

—— "The Chronology of Organized Sport in Nineteenth Century Scotland: A Regional Study II – Causes," *IJHS* 7.3 (December, 1990), pp.365–387.

—— "The Social and Occupational Structure of Organized Sport in Central Scotland during the Nineteenth Century," *IJHS* 4.3 (December, 1987), pp.301–314.

Wigglesworth, N., "A History of Rowing in the North West of England," *British Journal of Sports History* 3.2 (May, 1986), pp.145–157.

Coal-miners

General

Benson, J., *British Coalminers in the Nineteenth Century: A Social History* (Dublin, 1980).

Douglas, D. and Krieger, J., *A Miner's Life* (London, 1983).

Flinn, M.W., *The History of the British Coal Industry, Vol. 2, 1700–1830: The Industrial Revolution* (Oxford, 1985).

John, A.V., *By The Sweat of Their Brow: Women Workers at Victorian Coal Mines* (London, 1984).

Mitchell, B.R., *Economic Development of the British Coal Industry, 1800–1914* (Cambridge, 1984).

Pollard, M., *The Hardest Work Under Heaven* (London, 1984).

Simpson, J.B., *Capital and Labour in Coal Mining During the Past Two Hundred Years* (Newcastle, 1900).

North East

Challinor, R. and Ripley, B., *The Miners' Association: A Trade Union in the Age of the Chartists* (Whitley Bay, 1968).

Colls, R., *The Collier's Rant: Song and Culture in the Industrial Village* (London, 1977).

—— *The Pitmen of The Northern Coalfield: Work, Culture, and Protest, 1790–1850* (Manchester, 1987).

Colls, R. and Lancaster, B. (eds), *Geordies: Roots of Regionalism* (Edinburgh, 1992).

Fynes, R., *The Miners of Northumberland and Durham* (Newcastle, 1986).

Hair, P.E.H. (ed.), *Coals On Rails Or The Reason Of My Wrighting: The Autobiography of Anthony Errington from 1778 to around 1825* (Liverpool, 1988).

Hudson, M., *Coming Back Brockens. A Year in a Mining Village* (London, 1995).

Kirkup, M., *The Biggest Mining Village in the World: A Social History of Ashington* (Newcastle, 1993).

Lawson, J., *A Man's Life* (London, 1946).

Leifchild, J.R., *Our Coal and Our Coal Pits* (London, 1968) (first published in 1853).

Louis, H., *The Pitman's Yearly Bond* (Newcastle, 1930).

McCullough-Thew, L., *A Tune for Bears to Dance To* (Berwick, 1992).

—— *The Pit Village and The Store* (London, 1985).

Martin, S.B., *Barrington Colliery* (Bedlington, 1979).

—— *Bedlington* (Bedlington, 1986).

—— *Bomarsund and Stakeford* (Bedlington, 1987).

—— *Netherton (Nederton)* (Bedlington, n.d.).

—— *Sleekburn (The Station) with Bank Top, East Sleekburn and Red Row* (Bedlington, n.d.).

Moore, R., *Pit-Men, Preachers and Politics. The Effects of Methodism in a Durham Mining Community* (Cambridge, 1974).

Murray, A.V., *A Northumbrian Methodist Childhood*, edited by G.E. Milburn (Morpeth, 1992).

Patterson, W.M., *Northern Primitive Methodism* (London, 1909).

Redmayne, Sir. R., *Men, Miners, and Memories* (London, 1982).

Robinson, J., *Francie* (Newcastle, 1997).

—— *Tommy Turnbull. A Miner's Life* (Newcastle, 1997).

Welbourne, E., *The Miners' Unions of Northumberland and Durham* (Cambridge, 1923).

Williamson, B., *Class, Culture and Community: A Biographical Study of Social Change in Mining* (London, 1982).

Other

North East

Baldwin, C.E., *The History and Development of the Port of Blyth* (Newcastle, 1929).

Balmer, R., *A List of Nineteenth Century Licensed Premises in the Blyth, Newsham, Cowpen and Bebside Townships* (Blyth, 1974).

—— *Blyth in Old Picture Postcards* (Zaltbommel, Netherlands, 1983).

Bibby, R., *Bothal Observed* (Newcastle, 1973).

Craster, H.H.E., *A History of Northumberland*, Vol. VIII (Newcastle, 1907).

—— *A History of Northumberland*, Vol. IX (Newcastle, 1909).

Gerrard. R.E., *Northern Notabilities* (Newcastle, 1897).

Greaves, *Foxhunting in Northumberland* (Newcastle, 1965).

Harbottle, G., *A Century of Cricket in South Northumberland, 1864–1869* (Newcastle, 1969).

Harrison, W., *Ashington in Old Picture Postcards* (Zaltbommel, Netherlands, 1983).

Hodgson, J., *A History of Northumberland*, Vol. II (Newcastle, 1832).

Mackenzie, E., *Descriptive and Historical Account of the Town and County of Newcastle-Upon-Tyne*, 2 vols (Newcastle, 1927).

—— *History of Northumberland*, 2nd edition (Newcastle, 1825).

McCord, N., *North East England. The Region's Development, 1760–1960* (London, 1979).

Middlebrook, S., *Newcastle-Upon-Tyne: Its Growth and Achievement* (Newcastle, 1950).

Milburn, G.E. (ed.), *A. Victor Murray. A Northumbrian Childhood* (Northumberland Library, 1992).

Moffatt, F.C., *A Tap at the End of the Raw* Morpeth, n.d.

—— *For a Purse of Gold* (Morpeth, 1986).

—— *Turnpike Road to Tartan Track: The Story of Northern Foot Handicaps* (Morpeth, 1979).

Pike, W.T. (ed.), *Contemporary Biographies* (Edinburgh, 1985).

Ridley, Viscountess (ed.), *Cecilia: The Life and Letters of Cecilia Ridley, 1819–1845* (London, 1958).

Wells, J.A., *The Blyth and Tyne*, Part 1 (Morpeth, 1989).

—— *The Blyth and Tyne Branch, 1874–1989* (Morpeth, 1990).

Working class

Benson, J., *The Working Class in Britain, 1850–1939* (Essex, 1989).

—— (ed.), *The Working Class in England, 1875–1914* (London, 1985).

Harrison, J.F.C., *The Common People. A History from the Norman Conquest to the Present* (London, 1984).

Hobsbawn, E., *Worlds of Labour. Further Studies in the History of Labour* (London, 1984).

Joyce, P., *Visions of the People* (Cambridge, 1991).

—— *Work, Society and Politics: The Culture of the Factory in Later Victorian England* (London, 1982).

Meacham, S., *A Life Apart: The English Working Class, 1890–1914* (London, 1977).

Thompson, E.P., *The Making of the English Working Class* (Vintage, 1963).

—— *Methodism and the Working Class Movements, 1800–1850* (London, 1947).

Wearmouth, R.F., *Methodism and the Struggles of the Working Classes, 1850–1900* (Leicester, 1954).

General

Barker, T.C. and J.R. Harris, *A Merseyside Town in the Industrial Revolution: St. Helens, 1750–1900* (London, 1959).

Barnard, H.C., *A Short History of English Education, from 1760 to 1944* (London, 1947).

Briggs, A., *Victorian Cities* (Middlesex, 1968).

Cannadine, D. and Reeder, D. (eds), *Exploring the Urban Past. Essays In Urban History by H.J. Dyos* (Cambridge, 1982).

Church, R.A., *Economic and Social Change in a Midland Town, Victorian Nottingham, 1815–1900* (London, 1966).
Dennis, R., *English Industrial Cities of the Nineteenth Century: A Social Geography* (Cambridge, 1984).
Donajgrodzi, A.P., (ed.), *Social Control in Nineteenth Century Britain* (London, 1977).
Dyos, H.J., *Victorian Suburb: A Study of the Growth of Camberwell* (Leicester, 1966).
Harloe, M., *Swindon: A Town in Transition* (Heinemann, 1975).
Harrison, B., *Drink and the Victorians: The Temperance Question in England, 1815–1872* (Pittsburgh, 1971).
Hoggart, R., *The Uses of Literacy* (Chatto & Windus, 1957).
Jones, G.S., *Outcast London: A Study in the Relationship Between Classes in Victorian Society* (London, 1984).
Joyce, P., *Visions of the People* (Cambridge, 1991).
—— *Work, Society and Politics: The Culture of the Factory in Late Victorian England* (London, 1982).
Leighton, M., *Men at Work* (Jill Norman Ltd., 1981).
Mingay, C.E. (ed.), *The Victorian Countryside*, Vol. 2 (London, 1981).
Stone, L. and Stone, J.C.F., *An Open Elite? England, 1540–1880* (Oxford, 1984).
Thompson, E.P. *Customs in Common* (London, 1993).
Thompson, F.M.L., *English Landed Society in the Nineteenth Century* (London, 1963).
—— (ed.), *The Cambridge Social History of Britain, 1750–1950*, Vol. 2 (Cambridge, 1990).
—— (ed.), *The Cambridge Social History of Britain, 1750–1950*, Vol. 3 (Cambridge, 1990).
Wilson, J., *Memories of a Labour Leader* (Sussex, 1980).
Yeo, E. and Yeo, S. (eds), *Popular Culture and Class Conflict, 1590–1914* (Sussex, 1981).

Government

Clarke, J.J., *A History of Local Government of the United Kingdom* (Westport, 1978).
Laski, H.J, Jennings, W.I. and Robson, W.A. (eds), *A Century of Municipal Progress. The Last Hundred Years* (London, 1976).
Osborne, B., *Justices of the Peace, 1361–1848. A History of Our Magistracy During Five Centuries* (Dorset, 1960).
Smellie, K.B., *A History of Local Government* (London, 1963).
Taylor, J.M., *England's Border Country: A History of Northumberland County Council, 1889–1989* (Morpeth, 1989).

Theoretical

Bocock, R., *Hegemony* (London, 1986).
Bottomore, T.B. and Rubel, M., *Karl Marx. Selected Writings in Sociology and Social Philosophy* (Middlesex, 1970).
Clarke, J., Critcher, C. and Johnson R. (eds) *Working-Class Culture. Studies in History and Theory* (St. Martins Press, 1980).
Feuer, L.S., *Marx and Engels: Basic Writings on Politics and Philosophy* (New York, 1959).
Geertz, C., *The Interpretation of Cultures* (Basic Books, 1973).
Giddens, A., *The Constitution of Society* (Berkeley, 1984).
Guttmann, A., *From Ritual to Record: The Nature of Modern Sports* (New York, 1978).
Hall, S. and Jefferson, T. (eds), *Resistance Through Rituals. Youth Subcultures in Post-war Britain* (London, 1976).
Jenks, C. (ed.), *Cultural Reproduction* (London, 1993).

Kern, S., *The Culture of Time and Space, 1880–1918* (Cambridge, Mass., 1983).
Lukes, S. (ed.), *Power* (Oxford, 1987).
Samuel, R., *Peoples's History and Socialist Theory* (London, 1981).
Williams, R., *Culture* (Glasgow, 1981).
—— *Marxism and Literature* (Oxford University Press, 1977).
—— *Culture and Society* (Middlesex, 1966).
—— *The Long Revolution* (New York, 1966).

Theses and dissertations

Dent, L.M., "The Influence of the Coal Mining Industry on the Geography of the Ashington–Bedlington Area," Unpublished M.Sc. thesis, University of Durham, 1957.
Gordon, A., "The Economic and Social Development of Ashington – A Case Study of a Coal-Mining Community," Unpublished M.Com. thesis, 1954.
Hitchcock, J., "Belsay: A Study of a Northumbrian Country Estate," Senior Honours Dissertation, University of St Andrews, 1987.
Parratt, C.M., "With Little Means or Time: Working-class Women and Leisure in Late Victorian and Edwardian England," Unpublished Ph.d. dissertation, Ohio State University, 1994.
Seeley, J.Y.E., "Coal Mining Villagers of Northumberland and Durham. A Study of Sanitary Conditions and Social Facilities, 1870–1880," Unpublished M.A. thesis, University of Newcastle-Upon-Tyne, 1973.

Index

eBooks – at www.eBookstore.tandf.co.uk

A library at your fingertips!

eBooks are electronic versions of printed books. You can store them on your PC/laptop or browse them online.

They have advantages for anyone needing rapid access to a wide variety of published, copyright information.

eBooks can help your research by enabling you to bookmark chapters, annotate text and use instant searches to find specific words or phrases. Several eBook files would fit on even a small laptop or PDA.

NEW: Save money by eSubscribing: cheap, online access to any eBook for as long as you need it.

Annual subscription packages

We now offer special low-cost bulk subscriptions to packages of eBooks in certain subject areas. These are available to libraries or to individuals.

For more information please contact webmaster.ebooks@tandf.co.uk

We're continually developing the eBook concept, so keep up to date by visiting the website.

www.eBookstore.tandf.co.uk